ENVIRONMENT, EVOLUTION AND VALUES

About the Author

D.P. Chattopadhyaya, MA., LL.B., Ph.D. (Calcutta & London School of Economics), D.Litt. (Hony.), was Professor of Philosophy (Jadavpur University) Researched, taught and lectured at many Universities and Research Institutes in Asia, Europe and the USA (1954-1994). Chairman, Indian Council of Philosophical Research (1981-90): Chairman Indian Institute of Advanced Studies, Shimla (1984-91). Held public offices like Union Cabinet Ministership and State Governorship.

He published 35 books of which 18 are authored and 17 edited. Among his notable publications are *Individuals and Societies* (1967), *Individual and Worlds* (1976), *Anthropology and Historiography of Science* (1990), *Karl Marx and Sri Aurobindo* (1988), *Induction Probability and Skepticism* (1991), *Sociology, Ideology and Utopia* (1997), *Phenomenology, Philosophy and Other Essays* (2003), *Self, Society and* Science (2005), *Religion, Philosophy and Science* (2006).

Currently Chattopadhyaya is Director and Editoı of 96-Volume Interdisciplinary Project, *History of Science, Philosophy and Culture in Indian Civilization* [PHISPC]. He is a life Member of Russian Academy of Science (Moscow), Member of Intenational Institute of Philosophy (Paris).

ENVIRONMENT EVOLUTION AND VALUES

D. P. Chattopadhyaya

CONCEPT PUBLISHING COMPANY, NEW DELHI-110059

Cataloging in Publication Data—DK
Courtesy: D.K. Agencies (P) Ltd. <docinfo@dkagencies.com>

Chattopadhyaya, Debiprasad, 1933-
Environment, evolution and values / D.P. Chattopadhyaya.
p. cm.
Includes bibliographical references.
Includes index.
ISBN 818069366X

1. Historicism. 2. Human evolution. 3. Human ecology. 4. Social ecology. 5. Values.
I. Title.

DDC 901 22

ISBN : 81-8069-366-X

First Edition 2007

Published and Printed by
Ashok Kumar Mittal
Concept Publishing Company
A/15-16, Commercial Block, Mohan Garden
NEW DELHI-110059 (India)

Phones : 25351460, 25351794
Fax : 091-11-25357103
Email : publishing@conceptpub.com

To My Teachers

Sir Karl R. Popper	(in memory)
Professor A.C. Das	(in memory)
Professor J.W.N. Watkins	(in memory)
Professor K.K. Banerjee	(in memory)

in esteem, affection and gratitude

Preface

Of the fifteen papers included in the book, 3, 7, 12 and 13 have been published before : 3 in my book *History, Society* and *Polity*, Macmillan, New Delhi, 1976; 7 in *Philosophy : Theory and Action,* Poona, 1980; 12 in *Philosophy : Theory and Action,* Poona, 1980; and also in *Philosophy and Social Action* Vol. I, No. 2; 13 in *Philosophy of Human Rights Actions* Vol. I, No. 2, 1975 and also in *Philosophy of Human Rights* (Alan S. Rosenbaum ed.), Greenwood Press, West Port, Connecticut, 1980.

All other papers, except 2 and 9 have been presented before some or other conference, seminar, symposium or learned society. The first essay was presented at a Seminar on Man held at the Centre of Advanced Studies, Shimla, in 1973; 4 at the Pierre Teilhard de Chardin Birth Centenary symposium at UNESCO, Paris, in September, 1981 (revised later on); 5 at Sri Aurobindo Bhavan, Calcutta, in December, 1981 (revised later on); 6 at Interdisciplinary Seminar on Man and Nature at University of Poona in January, 1981; 8 at a Seminar on Philosophy in Science at Jadavpur University in February, 1980; 10 at the Golden Jubilee Session of the Indian Philosophical Congress. Delhi, in December, 1975; 11 at Gandhi Study Centre, Jadavpur University, in March, 1974; 12 at All Indian Sociological Conference, Banaras, 1975, and the present version before the Centre for Asian Affairs at MIT, Cambridge in March, 1982; and 14 at Seminar on Politics and Morality at Rajasthan University, Jaipur in June, 1980.

I wish to thank the Editors and Publisher of the Publications listed above for permission to use the materials in this book. The papers have been organized into Parts according to subject, and a new introduction has been written for each part. They are intended to provide conceptual notes concerning the problems dealt with in the papers, to relate the papers to each other, to other papers in the book, and to the current issues delineated and debated by other investigators and to carry investigation a little beyond. For advice, encouragement, comment and criticism I am indebted to many teachers, students, colleagues and friends, especially K. K. Banerjee, J.W.N. Watkins, H.D. Lewis, P.K. Sen, Kalyan Sengupta, T.K. Sarkar, P.K. Mukhopadhyaya, Archana Banerjee, Tirthanath Banerjee, Krishna Roy, and Shefali Moitra. It is a great pleasure to acknowledge my obligations to Professor S.P. Banerjee and Dr. (Mrs) Minakshi Roy Choudhuri, who read most of the text; and to Professors Ernest Gellner, Erwin N. Hiebert, Hilary Putnam, Thomas Kuhn, Daya Krishna, Kireet Joshi, and Chhanda Gupta, each of whom read one or more chapters, None of them is responsible for the mistakes in the book, especially since their advice was not always heeded. The influence of the writings of Marx, Sri Aurobindo and Popper should be evident. I have benefited as much from the ideas which I have criticized as by those which appear closer to my own. It would be clear from the following pages.

It is only during the first three months of 1982 I could manage uninterrupted time to complete the work necessary for converting the essays, though otherwise thematically related, into a book. My work has been supported partly by a Fulbright Scholarship enabling me to visit and work at Harvard University. Jadavpur University granted me study leave for nine months. I am grateful to the concerned authorities.

I would like to thank my friends, Buddhadev Bhattacharya and Vinod Kumar, for taking sustained interest in the production of this book.

Jadavpur University
Calcutta 700032

D.P. Chattopadhyaya

Contents

Introduction to the Second Edition

Long back, nearly a quarter of a century back, the first edition of this book was published. For more than a decade it has been out of print. Because of my preoccupation with other works and my previous Publisher's change of the area of publication, I could not bring out this second edition requested by many. And for which I must extend my apology to the interested inquirers and readers.

Called upon to bring out this second edition when I was reflecting on the main contents and arguments of the book, I felt the literature on the subject has become so vast since the time it first appeared that if I try to update the book in a substantial manner, I will be obliged to write a new book. Unfortunately, I do not have the necessary time or academic preparation for the purpose.

Both theoretically and experimentally, the basic concepts on evolution and their various implications —axiological, theological and sociological have been undergoing very significant changes. But, philosophically speaking, one can always revisit the old ideas with profitable insights provided by the *modern* scholarship. Reflective or philosophical theory of evolution does not get lost in experimentally and newly found details. The very possibility of reconstructing a coherent theory of evolution is strongly attested by the numerous works which have appeared during the last 25 years in this area of research and reflection.

The insights contained in the classic works of Lamarck (1744-1829), Charles Darwin (1809-1882), Gregor Johann Mendel (1822-1884), Hugo Marie De Vries (1848-1935) and few others have been extensively appropriated, significantly modified and updated by the famous life scientists like Theodosius Dobzhansky, G. G. Simpson, René Dubos and Ernst Mayr and their followers. Considerable philosophical depth has been imparted to this debate by the works, among others, of Karl Popper, John Eccles, E. O. Wilson, Z. Pylyshyn and Philip Kitcher.

The central point of the debate may be represented, theoretically speaking, by a simple question like : is there any antagonism between Nature and Nurture, between the realm of causal compulsion and that of creative freedom? I am afraid, even this question is perhaps faulty.

To many, this simple formulation of an otherwise very complex question may appear an oversimplification. Particularly from the hard-core experimental scientists and their philosophical interpreters, this reaction is sure to come. But I am always reminded of the fundamental methodological principle that whenever we are called upon to discuss fruitfully a complex issue, we must distinguish the *essentials* from the *non-essentials* and try to understand the latter in terms of the former. Otherwise, we are doomed to land in an area of fallacies like infinite regress, vicious circle or *petitio principii.*

Instead of discussing the issues at stake in an abstract manner, if we try to bring the concerned principles to bear upon the perceptually insistent facts, not necessarily experimentally measurable data, we would be considerably benefited and our questions too would get significantly clarified. In no area of science or science-rooted philosophy one can rationally expect to reach rock-bottom certainty or a permanent terminal point. Added clarification and increasing precisification are all that we can reasonably expect. I do not find any valid argument to dramatize the distinction between "normal science" and "revolutionary science".

In this connection the names, those of the "greats" of the last two millennia, which readily come to my mind are Caraka of *Caraka-samhita* fame, Hippocrates, Galen and Avicenna (or Ibn Sina). From their works available with us it is found that they could easily commute between facts and theories — facts to theories and theories to facts. They developed what is known today a sort of bootstrap theory. In the European tradition from Hobbes to Marx we find a sustained philosophical effort to tame or regulate philosophical theories by empirical facts. It may be recalled in this connection that Marx himself was an admirer, at least to a certain extent, of Darwinianism.

I refer to these thinkers mainly to illustrate one simple point: their naturalism or materialism was formulated in a way which could be shown, without strain, to be consistent with human freedom. A substantially similar point is evident in the political and philosophical writings of Spinoza. That partly explains why his life and thought are described in terms of seemingly incompatible expressions as "divinely intoxicated" and "materialist". To these thinkers *necessity* and *freedom* did not appear antithetical position. In the context of evolution their views do not encounter the so-called problem of "evolution *versus* creation". In their schemes of thought, emergence and creativity are not incompatible with natural necessity which science aims to discover.

On the questions whether human beings, extant and extinct, have all come from the zoological family, *Hominidae,* and whether one particular species, *Homo sapiens*, evolved out of it, the experts are not always unanimous. Though it is generally claimed that *Homo sapiens* experimentally appear to be descendents of African apes, which are believed to have lived 5-6 million years ago, but all professionals do not endorse this view. As in paleoanthropology and historical linguistics, the theory of diffusion, single point of origin, and subsequently, has been repeatedly disputed. And as an alternative to it the theory of parallelism, i.e., parallel growth, has been advanced. But if the hypothesis of single origin is insisted upon, then many problems which arise cannot be answered or, at any rate, cannot be satisfactorily answered. The alternative hypothesis is that human origin, origin of languages and origin of cultures are traceable to different parts of the earth. The emergence, adoption and diversification of primates and humans, it has been claimed, started in different parts of the world like Africa, Austronesia, China and Eurasia. The ancestral stock of extant African apes remains unknown. It seems the oldest known *Hominidae* have been the extinct genus called Australopithecus. Southeast Asian tarsiers, the Madagascan lemurs, the South American monkeys and the African big apes like gibbons, orangutans, chimpanzees and gorillas may be differentiated species. No evidence, fossil or lithic, can conclusively establish with knock-out certainty that *Homo erectus* first appeared only at one particular place of the earth.

It may be recalled here that Darwin's *On the Origin of Species by Means of Natural Selection* to a considerable extent rests on his geological findings. The ultimate source of genetic variation is found to be changed in the genetic material itself—DNA—which contains the information which is hereditarily passed on to the offsprings from generation to generation. The DNA information is contained in and carried on through long strands known as chromosomes.

Differentiation, i.e., alteration of the genetic material, is due to mutation. The vast majority of these alterations lead to the death of the individual or to his reduced genetic fitness, leading to death later on. If genetic variation is the law, then random variation is only an extrinsic bye-law, disrupting genetic equilibrium. Every sort of populations exhibit two traits, evolutionary and environmental. Several mechanisms are at work to maintain a viable relationship between a particular population and its environment. On analysis it is found that what is called Natural Selection is both conservative and dynamic in its character. Climate and food are among the forces at work for bringing about a viable relationship between a population and its environment. To diversify, natural selection is believed to be one of the important processes by which evolutionary changes will take place. Blood types, abnormal haemoglobin and externally induced diseases substantially account for the process of diversification. Genetic drift is also to be recognized as one mechanism by which evolutionary changes may take place. Organic evolution takes place as a consequence of the process of Natural Selection influencing a pool of normal variability within a living population.

Natural Selection operates at several levels within a population. The behaviour of the concerned population depends much upon the reproductive advantage of the individuals contained in the concerned population. In the scale of altruism and egoism how a particular individual belonging to a particular population would behave is largely determined by its kinship structure, nearness or distance of the parents and grandparents from the descendants, agnatic or cognatic.

In the scale of evolution at the top stand primates, viz., monkeys, apes, apemen and humans. If primates form an order, anthropoid constitutes its suborder. Under that suborder we find *hominoidae,* a sort of superfamily, comprising apes, ape-men and humans. Under the superfamily is subsumed family comprising only ape-men and humans, excluding apes. Under that superfamily of *hominoidae,* we find family, *hominidae,* consisting only of ape-men and humans. Subfamily under family is called *homininae,* i.e. humans. From the subfamily evolves the genus called *Homo.* And from which evolved *Homo erectus* and *Homo sapiens.*

The *Homo sapiens* populations of the Southern human range during most of the Pleistocene epoch and of a Neanderthal stage of human evolution are all now extinct. Having realized that the Neanderthals appeared both in the Southern human range and also in the Northern human range, comprising ancestors of a unique Eurasian race, evolutionists and physical anthropologists have developed the hypothesis of *Homo sapiens.* They are taken to be the descendents of local *Homo erectus* populations.

Many professionals maintain that they are the direct ancestors of the relatively recent and living populations of East Asia and Australasia. The latest remains of *Homo erectus* have been found both in China *(Zhoukoudian,* s.c., 3,70,000 years ago) and India (Narmada valley in Central India, around the same time). In the South, a large sample of early Australasian *Homo sapiens* has been found in Indonesia (the Ngandong site in Eastern Java). Several sites in China, India and Java have been identified where *homo erectus* and *homo sapiens* species had appeared.

It is possible that the *Homo sapiens* of China, Java and Australia are somehow related. But diffusionists are not keen to try to establish their migratory/immigratory links. It must be stated here that dating the fossils, even after the invention of the new technology of carbon dating, has been proving somewhat tentative, requiring periodic revision. It has been evidently clear from the varying dates of the African Stone Age over a period of 100 years or so. Conceptions of the course of evolution of the *Homo sapiens* in different parts of the world have changed drastically since the 1960s. The hypothesis of diffusionism has received added confirmation by the gradually accepted practice of *geographic separation* of human groups. Geographic races may be numerically large or small. Some of these smaller groups are found to be reproductive isolates.

The concept of demographic isolates is likely to remind one of the linguistic isolates like Basque (France-Spain border), Brahui (Baluchistan in Pakistan), Burushaski (near Gilgit in the Hindukus Range, Pakistan) and Ket (North-East Siberia). Both demographic-reproductive and linguistic isolates have naturally drawn the attention of concerned scholars, giving rise to considerable literature on the subject.

Both in Europe [e.g., Bilzingsleben in Germany] and North Africa [e.g., Sale, Rabat and Temara in Morocco] sites have been found of *Homo erectus* in the antiquity. The sites of *Homo sapiens* are numerous both in Europe [e.g., Paviland, Cro-Magnon and Krapina] and North Africa [e.g., Dar-es-Soltane, Jebel, lrhoud and Thomas Quarries]. The Moroccan sites of *Homo erectus* could not be convincingly related to the sub-Saharan sites of Africa. The classificatory principles used for identifying the sites of *Homo erectus* and those of *Homo sapiens* have been found to be somewhat fuzzy or inexact: and for that reason some sites like Mauer,

Vertesszollos and Aerago could not be definitely classified. From all these findings and attending uncertainty it would be inadvisable to maintain that the origin of *Homo erectus* and from which *Homo-sapiens* evolved could also be traced to any single place of the earth.

During the past 200 years, particularly following the Darwinian revolution, the theories of human evolution, adaptation and differentiation witnessed steady development. This was possible mainly because of his pro-naturalistic or scientific approach, dispensing with the traditional theory of evolution according to design defended, among others, by theologicians from Thomas Acquinas to William Paley. Darwin found that evolutionary change through time and evolutionary diversification, leading to multiplication of species, are not directly due to natural selection. They often come about as bye-products of natural selection as it fosters adaptation to different environments.

Besides Darwin and independent of him, another scientist, Alfred Russell Wallace [1823-1913], developed the concept of natural selection in 1858 and sent it to Darwin for his professional perusal. Herbert Spencer [1820-1903], a younger contemporary of Darwin and a professional philosopher and sociologist, also wrote on the thesis of the survival of the fittest, presumed to be a corollary of natural selection. A pernicious but unintended consequence of the theory of natural selection emerged from careless extension of the idea of the struggle for existence to economic and social life. Social Darwinism, in the name of popularizing the ideas of Darwin, at times vulgarized it. Social Darwinians like Chamberlain and Benjamin Kidd used Darwin's ideas to promote the theory of social struggle in a mindless manner, objectively encouraging brutality and suppression. Many thinkers relate the rise of Nazism and Fascism to the vulgarization of social Darwinism.

It is true that both Marx and Engels appreciated and endorsed some parts of Darwin's view which appear to them consistent with their own views. But both of them, particularly Marx himself, were conscious of the fact that the analogy between their views breaks down beyond a point. In fact, Darwin's idea of the survival of the fittest were not acceptable to Marx without substantial modification.

The person who should be credited to have provided the missing links to the Darwinian theory of evolution is an Augustinian monk Gregor Mendel [1822-1884] who, familiar with Darwin's *Origin of Species,* started a long series of experiments with peas in his monastery garden in Brunn. Because of his withdrawn nature and inability to publish the results of his experiments in any famous scientific journal of the time, his quasi-revolutionary findings pertaining to the principles of genetics did not receive due attention. The relatively obscure local journal in which he published his findings did not reach his professional peers. When, later on, the statistical analysis of his findings was known, it significantly modified and corrected Darwinian view of heredity. It was mainly due to the works of Hugo de Vries [1848-1935], a Dutch botanist, that Mendel's view started receiving its overdue recognition. De Vries found the great flaw in Darwin's view that there was no explanation in it for the way in which individuals may vary within a species. Yet it was only on that unexplained means of variation that the changes of evolution could be explained. De Vries developed his view accounting for different characteristics which might vary independently of each other and come together in different combinations. In fact, this formulation of De Vries was a rediscovery of Mendel's theory of genetics. In effect, De Vries was able to go beyond Mendel in some respects. He propounded a new doctrine of evolution by sudden jumps of mutations. De Vries's rediscovery of the Mendelian law of inheritance and adding to them his doctrine of mutation substantially improved Darwinian theory and successfully completed its promised structure.

Darwin's own faulty formulations encouraged the widespread tautology, *viz.*, (1) the fittest survive implies and is implied by (1A) the survived is the fittest. This empty tautology ignores the *indefinite* possibilities of mutation of *individuals* of a biological species or geographic species for various known and, particularly, unknown causes. So long as their causes and the probable processes of their effectuation remain unknown, any prediction regarding the evolutionary future of the species as aggregate continues to remain highly uncertain. If the evolutionary future of the species *qua* species remains uncertain, then the future of the *individual* members or sub-species under the same species continues to remain even more uncertain. The element of predictability, as we know, is an important component of the scientific claim of every scientific hypothesis.

In the area of life science there are different levels —human and (aggregatively speaking) sub-human. At each of these levels certain sublevels are discernible. The uncertainty of the predictability pattern increases or decreases depending on how the scientist approaches the evolutionary process —whether he does it according to an ascending scale or a descending scale. It is significant to recall that life scientists like Linnaeus (1707-1778), Buffon (1707-1788), Lamarck (1744-1829), Cuvier (1769-1832) and Lyell (1797-1875) had more or less familiarity with Natural Sciences like Geology and Physical Geography. Darwin himself too had a strong geological streak in his early scientific temper. Many of these life scientists in their professional career, to start with, had been interested in the details of classification and the principles of taxonomy. With the passage of time and accumulation of knowledge and its interpretation, the scientist's primary attention started shifting from classification to causation —causation marked by obvious variation and speciation.

In this connection, the works of Lamarck deserve special mention. He first engaged himself in formulating a natural method of classification for the vegetable kingdom. His work was like that of a naturalist philosopher who noted the important influence of environment, especially climate on vegetable development. From botany he moved to zoology; from zoology to chemistry and then to meteorology. Later on, he studied the disciplines of invertebrate zoology and paleoontology. Subsequently, before concentrating on evolution he also studied geology. His conclusions on evolution and their difference from those of Darwin have been found extremely influential and also controversial from 19th century onward.

One very important area of controversy relates to the issues of affinity and/or difference between the ideas of Darwin and Marx. In the light of Lenin's observations and the Soviet biology this controversy became additionally acute. In some of the official publications of the former USSR, the Darwinian view has been referred to as the materialist doctrine of the historical development of the living nature. Moreover, it has been claimed that by this doctrine Darwin dealt a fatal blow to metaphysics, theology and idealism. It is clear that this strong statement reflects not only pro-Darwinism but also its ideological overtone. Lysenko (1898-1976), a leading Soviet life-scientist, who wrote a very influential book on *Agribiology*, taking his cue directly from the writings of Michurin and basically from those of Darwin, advocated the view that Darwinism, unlike Lamarckism, lends its support to the Marxian view of social determinism. In this connection he coined the term "creative Darwinism". It must be added here that creative Darwinism has never been systematically presented by the ideologists on behalf of the former Soviet life-scientists. But Lysenko ascribes a clear creative role to selection and regards natural and artificial selection as the main factors in the morphological development of animals and plants. In Darwin's view selection is only the choice and not the generation of existing variations. But Lysenko changes the meaning of selection. He thinks that Darwinian natural selection involves joint operation of such factors as variation, heredity and survival [in place of over population]. To him, it may be noted, selection is the entire complex. It is by this definitional strategy that Lysenko and his followers developed an unorthodox view called "creative Darwinism". Lysenko's fame mainly rested on his political patronage. After the fall of Khrushchev, his pre-eminent position in the field of biology and agronomy declined and gradually disappeared.

Another area of controversy surrounding Darwinism is its social, rather than demographic impact. Darwin was influenced by Malthus. Marx was more influenced by Ricardo than by Malthus. But all the three and their followers favoured their differentiable theoretical positions and had to deal with such concepts as "struggle for existence", "competition" and "survival of the fittest". Many of them, particularly Spencer, were strong defenders of free competition. He tried to draw pro-competitive elements from the Darwinian source. Unfortunately, Darwin failed to abstract social relations convincingly from the socio-morphic component of his model, which encouraged reapplication of his views to different aspects of the society—ethical, social and political.

Even some biologists, for example, Pauling, raised the demand that in order to protect mankind's genetic stock, the State intervention is called for in the matter of choice of sexual partners. What is called welfare economics purported to improve the amelioration of social conditions was another fallout of the views critically aired at times against Darwinism. It may be mentioned here that as a professional scientist, Darwin was not particularly concerned about the direct ideological consequence of his views on *Theory of evolution*. At the same time, it must be added here that a versatile mind like his was not totally unconcerned with the possible social implications of his theory of evolution.

The area in which very consciously Darwin decided to swim against the contemporary current is theology. As alluded before, Darwin's teacher, Paley and many others, true to their pro-religious disposition, used to maintain that in the harmonious picture of nature as God's handiwork the case of man is assured by design. Darwin's rejection of the then widely accepted doctrine of creationism endeared him to Marxists.

Notwithstanding the Marxist or/and the materialist support to Darwinism, the development of the genetic research of the 20th century posed a genuine critique of Darwinism. Most of the newly appearing variations (mutations) which showed themselves to be hereditable reduced the vitality of genes, stood in no identifiable relationship to the triggering environment and could not be ordinarily reproduced. Most of the former Soviet biologists got closer to the view that the reactive norm of every living system is inscribed in a genetic code of the total structure of the germ-cell and not that of the DNA alone. If the changes brought about by external factors exceed the limits of the reactive norm, the concerned system would adopt a new homeostasis which gets genetically inscribed by a feedback of information to the coding system.

Another development which is evident in the Western research on the inheritance of the acquired characters is the increasing recognition of some aspects of Mendelian genetics. Also to be noted is the fact that in genetic research several Western writers like Ludwig, Thornley and Lindsey started paying more attention to Lamarckism in its updated form. The solution of this question touching the causes of genetic variability understandably captured the attention of the leading life scientists of the 20th century.

The modern synthesis, emerging out of the debates on the partly competing and partly overlapping views of the classical hypotheses of Lamarck, Darwin and Mendel, is yet to receive universal assent. In the area of experimental life sciences to expect any achievement like universal assent is perhaps contrary to the very spirit of science. Even then the peers in the area are expected to develop a sort of workable consensus.

It may be remembered that expressions like Lamarckism, Neo-Lamarckism, Darwinism, Neo-Darwinism, Mendelian genetics and Neo-Mendelian genetics are all outcome of different types of synthesis and interpretations offered by leading life scientists. In all these schools and sub-schools of thought one finds significant internal difference between the experts. For example, in the area of population genetics the views of J. B. S. Haldane (1932 and 1938), R. A. Fisher (1936), Sewall Wright (1968-78) and Mayr (1976 and 1982) exhibit both convergent and divergent views. The modern synthesis seems to be largely, but not entirely, based on what is called the alternative or balanced hypothesis of population structure. Their works have been influenced by those of Thomas Hunt Morgan (1915 and 1916), who in turn followed De Vries, and H. J. Muller (1973). Both Morgan and Muller are found to have rejected the important insights of the new population genetics, viz., a wide range of genes influencing each characteristic already exists in any natural population for selection to act upon. Genetic traits, due to mutation, continue to circulate within a given population at a low frequency even if they impart no advantage, and the species consequently build up a fund of variability that can be used by selection when conditions change. Even in a stable environment, balance-fostering selection effectively helps to maintain and regulate the range of genetic variation which subsequently can be exploited in adaptive evolution. Both Haldane and Fisher insightfully used mathematical language in their formulation on population genetics.

Later on, most field naturalists like Dobzhansky (1937 and 1977) and Provine (1971 and 1986) combined their own findings with mathematical insights provided by Fisher and Haldane. These enterprises to bridge the gap between the practical experience of the naturalists and abstract formulations of the experimental and mathematical biologists have yielded a rich harvest.

In the kindred area of human evolution the works of Rene Dubos (1965), Garland E. Allen (1978) and Eccles (1970 and 1976) shed new lights on human biology and proved very influential. Some of the scientists like Simpson (1944, 1955, 1965 and 1980) are strongly pro-positivist in their conclusions, while Eccles, like Karl Popper (1952 and 1974) and Pierre Teilhard de Chardin (1959) were not only anti-positivist but have a transcendental and even spiritual streak at times in their approach. Both Julian Huxley (1961 and 1963) and Chardin are strongly inclined to think that the process of natural selection, particularly the latter phases of evolution, are marked by a reogenetic, or hominizing tendency, culminating in the production of human mentality. It is not surprising that many life scientists like Medawar (1961 and 1967) are strongly critical of the hominizing claim of some life scientists.

Scholarly caution requires us to recall that within the framework of evolutionary hypothesis, the line of distinction, not division, separating the sphere of higher primates and humans need to be interpreted in terms of gradualism and not dualism. Even those who are thorough going naturalist can hardly deny, without some sort of inconsistency, that nature and nurture are not entirely separated. What is more, many animal psychologists try to show, on the basis of their observation and experiment, that many higher level animals, particularly primates, exhibit rudiments of intelligence in their behaviour and feelings. Some of them even recall selectively Darwin's observations in some of his writings (1873 and 1874).

From these pro-humanistic insights available in the evolutionary theory it will not be advisable to attribute excessive libertarian accent to biological evolution. These strong claims under evolutionary biology must satisfactorily account for our ethical concepts and make good sense of ethical judgements both about ourselves and others need to be taken with circumspection. Also it will not be easy to establish the view that evolutionary biology can teach us facts about human beings which, in conjunction with moral principles, can give rise to a comprehensive theory of justice and other fundamental concepts of moral philosophy. Equally strained seems to be the view that evolutionary biology is well-equipped to enable us to develop a set of meta-ethical principles commanding our acceptance.

In other words, those who try to burden evolutionary biology not only with finding facts and trends leading to emergence of nurture in the world of nature but also to elicit from it the higher principles of human unity and perfect global order, perhaps accept an undischargeable responsibility. In the area of human sociobiology, it is to be admitted, we badly need to use the insights provided by evolutionists, geneticists, psychologists, sociologists, anthropologists and cognitive scientists, but that does not warrant us to rush hurriedly to a pre-mature syncresis, not amounting to promising development. Also in this connection we should bear seriously in our mind the criticisms which have been and are still being advanced by those writers who are not hostile to evolutionary biology but firmly wedded to the principles of sound scientific method.

Philosophers of science of the Popperian persuasion, it has been stated, reject evolutionary biology on the alleged ground that it is unfalsifiable. This critical reference to Popper and incidentally to the evolutionary hypothesis itself needs to be cautiously evaluated. In fact, the history of falsificationism makes it abundantly clear that Popper (1962) himself substantially revised his early formulations of falsifiability. His later formulation (1934 and 1959) of the criterion of what is scientific is known as *testability*. Going further, specially in the context of speculative theories with empirical import, he suggested another kindred and considerably weaker criterion called *criticizability*. In fact, Popper in his later life considerably drew upon the Darwinian insights, sans its pro-deterministic suggestion, both in his epistemology and scientific methodology.

Popper's view on the subject is likely to remind one of the views of Hobbes and Hume. In a very important sense both of them were naturalist and at the same time strong defender of the ideal of freedom. It is well known that many writers referring to Hobbesian thought are found to forget his passionate defence of freedom in the sphere of state affairs. There is no incompatibility between the two. The idea of a strong state and the ideal of freedom can both be simultaneously defended. In this connection one is well advised to remember not only naturalists like Hobbes but also anti-naturalists like Hegel. In the Indian tradition one can point out the Samkhya view which affirms both naturalism and freedom at the same time. The plausibility of the simultaneous defence of naturalism and even antinaturalism, on the one hand, and human freedom and institutional freedom, on the other, can be worked out and shown by painstaking analysis.

A substantially similar problem reappears in the context of Hume's reference to the defence between *fact* and *value*. It is well known that Hume, together with Butler, Hutcheson, Cudworth and Adam Smith, was a prominent defender of the moral sense theory. In politics he was also a libertarian. To him that human action is *rational* which has a *traditional* sanction behind it. Closer analysis of his view shows that his psychology was double-layered, (i) distant-traditional and (ii) proximate-actual. If the desires and motives by which humans are induced to act are found to be proximate or immediate, we encounter one kind of psychological determinism. If, on the other hand, the desires and motives which from distant and traditional conditions, silted as sentiment and disposition, are found to be responsible for human actions, then those actions may be safely pronounced as free. The difference between the proximate psychological conditioning and distant psychological conditioning is a matter of degree and it is hardly possible to draw a clear line of distinction between them. Human consciousness or psycho-somatic awareness is found to move both backward and forward according to the inclination of the concerned human beings. In the light of these considerations, it seems, Humanism cannot be criticized as a defence of determinism simply on the basis of the isolated thesis of his so-called fact-value dualism. It would be incorrect to bracket him with the determinists of the naturalist persuasion. In this connection Watkins' elaborate discussion (1973 and 1999) of the views of both Hobbes and Hume are careful, perceptive and instructive.

Neither evolutionary biology nor quantum physics negates

individual or aggregative freedom. The exaggerated emphasis on the concepts like necessity, causation, determinism, freedom and their cognates are basically responsible for drawing the line of dichotomy between the world of nature and that of culture or nurture. The world comprising both nature and culture are more or less unifiable and yet perpetually open-ended.

December 10, 2005 **D. P. Chattopādhyaya**

Introduction to the first Edition

The essays of the book, though written on different occasions spread over eight years (1973-81), have an evident thematic unity and are concerned with the nature of man and his activities, both individual and collective. Different forms of knowledge and language-use are, on scrutiny, found to be basically social in character. In the book, a sustained attempt has been made to show the evolutionary nature of man and his activities, social and valuational. An equally sustained argument has been developed to show the inadequacy of the static-structural concepts of man and his activities. In the structural approach, the importance of time, evolution and history are systematically underestimated. There is a persistent impression among a section of structuralists that if *historism* and *evolutionism* are seriously recognized, then such concepts as *universality, necessity,* and *objectivity* cannot be satisfactorily accounted for. To my mind, the basic ideas of structuralism are associated with such names as Plato, Descartes, Kant, some neo-Kantians like Cassirer and Strawson, and Piaget, Levis Strauss and Chomsky.

It goes without saying that their views are not similar on all philosophical and social issues. But they have one point in common and that is denial of the evolutionary or the process character of reality and man's cognitive competence. In spite of his known sympathy for Plato, the thinker who made a most remarkable attempt to the process character of reality is to my mind, Whitehead. And it is a pity that I could not examine some of his arguments in the book. It is on this basic point that my difference with some structuralists would be evident in the following pages. Structuralism has been proved indeed very "seductive" and infectious. In contrast to them, I shall argue, among other things, that reality is evolving; the human ways of knowing reality are evolving; and human values are diverse and dynamic.

In course of my arguments, I would like to show that laws of science, though intended to be universal, are never strictly necessary in the cognitive sense of the term. Logical necessity is a postulation and not a cognitive achievement. Consistently with my realistic position I am obliged to believe that there are universal structural properties of reality, but since that obligation cannot be cognitively fully discharged, I cannot claim that human *beliefs* regarding those universal structural properties can be justified in any strong sense. By implication I admit that the universality claim of our knowledge should be subjected to critical and, if possible, experimental scrutiny. In fact, *history* of human achievements, social and cognitive, make this point evidently and abundantly clear. I have an insistent feeling that *history* of such *sciences* as cosmology and biology should be brought closer to philosophy and linguistics so that some of our basic and recurrent misconceptions are seriously and critically reviewed and removed.

When it is said that the human nature is evolving, it entails two other things. First, the concepts and theories in terms of which man organizes his sense-experiences are not innate and permanent but formed and subject to transformation. The symbols and the strings connecting those symbols are also found to be subject to the laws of social change. Secondly, the capacity underlying the formation and transformation of concepts and theories and symbols and strings of symbols purported to express the former is also subject to change. This *change in human capacity* is best understandable in terms of the process of *biological evolution* and *ecological interaction*. Our capacities and their exercise are contingent upon the availability or lack thereof some powers not entirely or permanently native to our nature.

There is no fixed *essence of* man which is insulated above the process of evolution and protected against the forces of ecological interaction. The essentialist or the structuralist fails to account for the change in Man's *linguistic competence*. Alternatively, he claims that the change is merely a surface-phenomenon, confined only to the level of *performance,* and does not affect the *basic* competence of man. From this insistent position of the structuralist arise a number of paired concepts: performance and competence; appearance and reality; phenomena and noumena; surface-structure and deep-structure; opinion and knowledge.

While as a gradualist I do not deny that there is a difference between the members of the said pairs, I am of the view that the difference is one of grade and not of kind. The thesis of gradualism that I propose to defend in the following pages is pitted against the sort of dualism that is there, implicitly or explicitly, in the ideas of Plato, Descartes, Kant and his modern followers, and the structuralists working in the fields of anthropology, psychology and linguistics. Realistic gradualism, defended in this book, clashes with cognitive and valuational universalism of Plato and Kant, structural universalism of Chomsky, and Piaget's psychological universalism. As regards Kant I would like to add, somewhat qualifying my general anti-structuralist position, that by redeploying some of the arguments, especially those pertaining to teleology, of the Third Critique Kant's view, unlike Descartes's one, can be brought closer to gradualism.

In fact I want to do away with man/nature dualism. Man is natural and simultaneously aware of his being natural. What we call human culture is humanly transformed or nurtured nature. Man's ways of transforming or appropriating nature are *partly* inherited from nature itself. Firm-foot in nature, man rises above nature, and becomes author of culture. It is through his concept forming and symbol-using capacities that individual men are in a position to exchange their experiences and ideas and pool them together for both individual and collective purposes. Though man is the basic *dynamo* of social dynamics, he cannot completely break away from the group dynamics of which he is both a creator and a creature.

Social change may be viewed from different aspects: synchronic and diachronic. Generally speaking, in cultural anthropology we come across a synchronic or static view of social change. As Bergson was fond of reminding us that a snap-shot (i.e., static) view of changing reality does not take away the changing character of reality itself. Rightly understood, the synchronic view of reality is not inconsistent with but rather complementary to, the diachronic or the historical and the evolutionary view of reality. While the cultural anthropologist concentrates on the details of social interaction within a given timeframe, the historian and the evolutionist tries to trace the successive forms of the human evolution, from the biological to the social stage. Interaction and change are two aspects of one and the same process, the same dialectical process. One might say: structure has history and history has structure. But, then, it has to be added: the structure of history has itself a history (of course at a higher level). The hierarchy does not end with "the Platonic Form of Forms" at its top.

The dialectic between man and nature, between nature and culture, both generates new needs in man and makes him increasingly conscious of the same. The human needs are partly natural, native to his biological being, complex extension of the powers and laws of nature, and partly social. The body of man is so equipped and organized that it can decode the encoded messages received from nature, interpret the same creatively, extrapolating beyond the given. It enables him to initiate actions, resist others actions, and adjust himself with different situations. What we call values are partly expressive of his needs, biological and social, and partly recognition of the properties of the objects he needs for his satisfaction of different sorts. The *valued properties* of the objects and *man's needs* of the same are neither fixed nor permanent: they interact, evolve and interact.

D.P. Chattopadhyaya

PART I

HUMAN EVOLUTION

HUMAN EVOLUTION

As specimens of very difficult questions about man one might ask, "what is the nature of man?" and "what is the essence of man?" It is not that these sorts of questions have not ever been asked. The answers to them, on examination, have been found to be as unsatisfactory as the questions themselves. Whatever answer one gives to either of the questions or both, viz., "man is a rational animal" or "man is a symbol-using being" or "man is the crown of creation", it is always possible to observe, critically, at the end, "is that all about man?" The impression that this type of observation conveys is that *enough* has not been said and that *more* could be said about man. The trouble is that there is no *end* to this quest for more. For it seems there is neither "*the* nature" nor "*the* essence" of man which can be *exhaustively* defined or stated at any particular point of time to the total satisfaction of the misled questioner. The main mistake of the questioner is that he himself does not know what answer would possibly satisfy him. And this is a pointer to the direction of how to raise questions rightly and where to look at for the right answers.

Essentialism fails both in science and philosophy, and it seems *evolutionism* is called for. Man is what he does and there is no end to the story of what he can possibly do. Man's history, like man himself, is open-ended. He is *not* the sum total of his doings. For that would be an unsatisfactory, superficial and nominalistic answer. What man wants to do, strives for, is also a part of his being. And it is mainly from man's doings and strivings that, Vico assures us, we can construct possibly the most intelligible account of what man is.

Without trying to enter into the elusive human "nature" or "essence" one can have a fairly clear, though not perfect, image of what man is from all that he does and, as a member of a species, or a community, has already done. Among many other things, man has succeeded, partly, in understanding and taming the forces of nature, both living and non-living, and in using symbolic language for the purposes of storing and communicating experiences and ideas. The implications of this success, though partial, are enormous and far-reaching. Pooling their informational sources together and making *needed* use of them, men have known their species, formed various social and other institutions, and, what is very important, become conscious of what they have been able to do and what remains yet to be done. This consciousness is historical, disclosing to man both his competence and performance, what he can possibly do and what he has done. Incidentally, this reflective consciousness reveals to man his dependence upon, relation with, and reinforcement by the laws of nature. Man knows that he is endowed with immense competence by nature and yet, at the same time, limited by its forces and subject, for example, to the laws of birth, growth, decay and death.

In the papers of part I of the work I have tried to focus my attention on some of the aspects of man's relation with nature on the one hand, and culture on the other. And it would be noted that I have tried to show the *graduated* or *continuous* character of the relation between the two. Man is a sort of *tertium quid* between nature and culture, endowed *with* the capacity to transform some parts of nature into objects of culture, and, as this capacity is finite and subject to the laws of wear and tear, he cannot do it in whatever way he pleases. It is only as *nature's natural* that man is the author of culture. Here 1 should clearly state, what is perhaps obvious from the context, that I am using the term man in the sense of *species-being* and not as *this* or *that* individual. This concept of man is not only intrinsically social but also *evolutionary* and *anthropological*.

Some eludications are in order. Nature-culture dualism or nature-nurture dualism, as it is sometimes said, can be done away with in several ways. Reality may be conceived of as *a unified* hierarchy of many ascending/descending levels—the physical, the mineral, the vegetative, the biological, the psychological and so on. The hierarchy may be considered as static-structural or dynamic-evolutionary, depending on the ontological status given to *time* in it. The unity of the different levels may be taken mechanically or teleologically (providentially). When man is deemed to be an evolute of a long evolutionary process one thing which is sought to be conveyed is this: he, though more free than other living evolutes, is finite, fallible, both in action and cognition, and perishable. These characteristics of man prompt me to characterize the view of man defended in the book as *anthropological* as distinguished from *metaphysical* or *essential.* In fact I am trying to describe what man is and not what his essence is, for it seems to me there is no *timeless essence* of man or for that matter even of species-man. The species-man, too, is subject to the laws of evolution as is evident from the researches in palaeontology, biology, neurophysiology, and other allied disciplines. Man is being continuously shaped and reshaped by what he receives from nature and culture, how he makes use of them, and gives back the "same" to their "sources". I say "the same" and "sources" because in and through the interactive process these also change their identifying characteristics.

Finitude, fallibility and perishability or mortality of man have, it seems, some important epistemological and cosmological implications. I have tried to clarify some basic issues concerning the scope and limits of human knowledge as influenced by the said characteristics in essays 2 and 3. It would be noted that essay has been included, without modification, from my previously published book, *History, Society and Polity* (Macmillan, New Delhi, 1976). My present studies, it may be observed, are the follow-up of what I have written and published earlier. Being finite and fallible as he is, man cannot know the world as such but only as interpreted by his body, mind, language and other attending conditions. In ultimate analysis man's is a body among bodies (of course with a privileged endowment and that is its reflective ability or self-consciousness) and its activities and propensities are, to a great extent, understandable in terms of the laws of natural and cultural dynamics. Body is both a limiting and an enabling condition of man's thought and action. By acting and thinking man keeps on changing his body, his self. For purposeful thought and action man needs and makes use of culturally available frame(s) of reference. Even for the purpose of creating a new frame of reference or theory, he uses an existing one, may be even an obsolete one. Therefore, both in thought and action, a continuity, evolutionary or historical continuity, is clearly discernible. A human creator is not a canvas-cleaner, does not create out of nothing. To know or to create man does not go out of himself, break away with his biological heritage and cultural tradition, and start everything *de novo*. Unless a frame of reference is accepted, even to the expression *"de novo"* a clear sense cannot be assigned. To break a frame one needs a frame. To break and make the frames necessary for thought and action, man is obliged to make use of the resources or competence biologically and socially made available to him. In a manner of speaking one can always say, as I have said, man cannot jump out of his own skin to think and act freely and fashion the world as he pleases. Nature and culture have a say even on his pleasure. Man has been nurtured by them both.

In essays 3, 4 and 5, I have tried referring critically to the views of Sri Aurobindo and Pierre Teilhard de Chardin, to trace in depth, the relation between species-man and nature in the broadest scientific sense, as studied by cosmologists. While I am deeply interested in the future of mankind, which partly explains my studies in the evolutionary theories of these two distinguished thinkers, I find certain difficulties in accepting their views on the *inevitable* evolution of the Kingdom of God on the earth. Given the validity of the Second Law of Thermodynamics and its ramifications, including the rule of entropy, I do not quite see how Sri Aurobindo's vision of the Supermind, and Teilhard's Omega-point could be defended. However, to my mind, their views and arguments are very insightful and interesting, and do deserve careful study. If my conclusions sound somewhat negative, it is because of my primary sympathy with science. The contemplation of the law-governed character of the universe, its origin and evolution, gives to many of us deep emotional satisfaction. Does one *need* an anthropomorphic God who can interface with the laws of nature or is himself bound by the same? If one does, is it not because of one's long cultural affiliation to some or other institutionalized religion? Some of these questions are bound to crop up in the mind.

Without knowing the universe he is in, man cannot be satisfied, rationally and emotionally. This knowledge, if available and seriously accepted, should not evoke in man either a myopic love for life or a rootless fear of death. It has been persuasively argued by scientists like Einstein and Eddington, who cannot be considered anti-religious, that the "doctrine of a personal God interfering with natural events could never be *refuted*... by science, for this doctrine can always take refuge in those domains in which scientific knowledge has not yet been able to set foot." Even Theodosius Dobzhansky, the distinguished biologist-philosopher, who does appreciate Teilhard's evolutionary ideas, has to admit that "such grand conceptions are patently undemonstrable by scientifically established facts."

1

Man : An Essay in Philosophical Anthropology

I
MAINLY ANTHROPOLOGICAL

Man is a multi-dimensional phenomenon. And as a phenomenon and true to the nature of a phenomenon, he is partly disclosed, mainly in the nature of potentiality. Since his disclosed being in main is physical-behavioural, it has to be reconstructed from cosmogenesis and physical anthropology. Man is said to be a noogenetic phenomenon, meaning thereby the gradual evolution of mind or mental property, and having a sub-mental legacy behind him.[1] The very idea of studying man in a cosmogenetic context, rather than in a cosmological one, is prompted by an obvious dynamic consideration. The human phenomenon can certainly be studied in a static cosmological or physical framework. But by following the hominization or evolutionary process perhaps we could have a clearer image of what man is and what possibly he can be.

The relevance of physiology and psychology to the study of physical man is indisputable. But while these two disciplines are mainly concerned with particular men and examine their individual mental acts and behaviours, anthropology focuses attention on men in groups, on races and peoples and their happenings and doings. There is something in the very nature of man or, one might say, in the very nature of the human body, which enables it to reflect on its own identity, recapture its changing identity of the past, and present its own future image to itself. While the cosmogenetic evolution is marked by diminishing simplicity of elements and increasing complexity of their permutations and combinations, curiously enough, the human evolution is characterized more by the process of convergence and less by that of divergence, This is perhaps the main reason for man's uniqueness in the scheme of natural phenomena. He is an abnormal phenomena in the normalcy of nature.[2]

Mainly Archaeological and Historical

Archaeological findings are fragments of the human past. They are disjointed and scattered bones of history. The physical past of man ordinarily lies behind him, but his historical past is in a way inalienable and exists within him. Mankind recognizes the objects of archaeological interests as excerpts of its own history. It is by his thought, his reflective ability that man can somehow fill in or at least narrow down the gaps between different and apparently unrelated ages of the past.[3]

History is concerned only with the human past. The temporal or the physical past falls beyond the pale of history. But the line of demarcation between the historical past and the temporal one cannot be sharply drawn. One hedges into the other. It is only from the reflective standpoint and by the deepening of thought that the demarcation between the two at any point of time can be somewhat provisionally indicated and not decided once for all.

Attempts have been made also to draw a line of demarcation between history and archaeology on the ground of the differences in their main sources. History is said to have been relying mainly on written records and, to a lesser extent, material relics; in the case of archaeological research, the material relics are primary and the written records are of secondary importance. On the basis of this untenable distinction it has often been argued that archaeology can hardly throw any light on the life of individuals or on single events and that in this respect its contribution to the understanding of the human past is somewhat of a limited nature. Once we realise that, except for theoretical purpose, the argument that social life and the individual life are absolutely inseparable from one another loses much of its force. When the archaeological objects are scientifically related in terms of local and regional stratigraphies and confirmed by the geologists and the climatologists, we get a clearer and chronological image of the evolving man. For analytical purpose, this chronological sequence may be reorganized as a typological series or defined as objects of the same class according to their formal similarity and dissimilarity.[4]

The social life is production of man's own life. Whether it is continuously and/or fully lived by man is a different matter. Even what we call material relics, rightly understood are cultural objects, and they are not to be confused with physical entities. Once the social identity of man and his cultural reference framework is more or less definitely indicated, the meaning of the material relics assumes an added contextual clarity. It is in this context that one has to understand the definition of archaeology as the anthropology of extinct cultures. The extinction of culture does not mean disappearance of all its cultural data. In fact by using specialised techniques the archaeologist obtains the cultural data of the human past, recaptures their meaning from the appropriate social contexts and enables the same to be used by anthropologists, historians, art critics, economists, and others interested in man and his activities.[5]

Mainly Linguistic

What mainly differentiates man from the rest of the animal world is his ability to signify and symbolize.[6] All human achievements, from propositions and speech acts to potteries and material relics, have an obvious common generic character. One might say, this is their symbolic character. Man's inalienable ability to comprehend objects, including sounds, stands in his way to blind *imitation* or faithful *reproduction. His comprehension transforms imitation into production.* It is only by introducing machine and relying on it that man can provisionally put a stop to his generative-productive capacity. Whatever man does is marked by an element of volition. Even his faithful reproduction or representation of his own experience is somewhat unfair to the facts of his own experience. This unfairness is not an act of design or a part of a plan. In spite of his organic ability and alertness, the human response to environmental stimulus is not exactly quantifiable. It is mainly for this reason that the radical theory of behaviourism miserably fails in the paradigm of symbolic activities, in the field of linguistic behaviour.[7]

The psycholinguist can certainly establish a statistical

correlation between linguistic behaviour and psychological phenomena, but that is not of fundamental interest. First, these statistical generalizations often encounter falsifying instances. Secondly, the reinforcing or motivating psychological phenomenon underlying a verbal behaviour is often found to have transformed, and not caused, the latter. In other words, the element of transformation has a role to play in between the psychological "cause" and the linguistic "effect". Psychology may influence but cannot cause man's choice of symbols, linguistic behaviour or speech acts. For similar considerations one might point out the inadequacy of causal explanation of ideology framed exclusively in terms of psychology.[8] The claim that scientific understanding of the human behaviour demands complete elimination of all uncertainties and insurance of correct prediction on the basis of environmental conditions is not only very narrow but also untenable. The behavioural system of a man cannot be neatly coordinated in a physical system. The explanatory coordinate of the verbal behaviour of man is particularly indefinite. The social (e.g. speaker-hearer) situation which adds flesh to the bones of speech acts are to be constructed in terms of flexible rules and regulations. Otherwise we will be obliged to condemn as meaningless many statements which are informative and/or successfully used for the purpose of communication.

Whether we like it or not, we cannot avoid reference to "internal states" like wills, impulses, feelings, purposes and so on. Rock's motion and man's emotion, for example, are not objects of the same category. Even if we follow the rigorous scientific method of physics in physiology and psychology, we cannot totally dispense with the notions of internal states. There is always an element of psychological alchemy accounting for the lack of one-to-one correspondence between the input and output of the human communication and information. The tests devised to confirm or contradict a particular hypothesis regarding the input-output relationship are themselves governed by certain subjective considerations. Computerization of information theory is not an answer to this irremediable subjective element in the matters of theory-construction and theory testing.[9]

Language may be studied both from the descriptive or

synchronic and the historical or diachronic points of view. It would be wrong to suggest that a historical structure is a succession of descriptive structures superimposed on one another and there is no distinction between the two. The descriptive approach is more suitable for the purpose of organizing and explaining a finite corpus, while the historical approach can take better care of irremediably fragmentary volume of data and the changing features of the same or similar data under different circumstances or situations. The descriptive or taxonomic study of linguistics fails to account for the transformative features of language.[10] That a language is indeed a form of life and that like all other forms of life it also undergoes silent but steady changes cannot be satisfactorily explained in terms of radical behaviourism or an archaic view of stimulus-response.

II
THE IDENTITY OF MAN

Man is not only a multi-dimensional being, he is also an ambiguous being. His multi-dimensionality and ambiguity are intimately interrelated. His consciousness is transparent but not his body. Even of his consciousness it has been said that, though in itself it is diaphanous, it is not represented so to us. The presentation of consciousness to itself is mediated by its bodily being. Whether the body itself is a form or idea of consciousness is a matter of a very old controversy. To counter the thesis of body-as-*form*-of-consciousness, or consciousness-as-*function*-of-body, has also been offered long ago; and its different formulations are popular in the pronaturalistic quarters, both Marxist and Positivist.[11] The other interesting view is being defended by the dualists who maintain that the principle of consciousness is altogether different from, and even alien to, the principle of body. They are somewhat puzzled by the "mystery" of the functional convergence, co-existence and what is more intriguing, cooperation of these two apparently alien principles in the identifiable locus of human personality. Interaction they understand; cooperation they cannot.

There are two extreme forms of the negation of dualism. In order to vindicate the primacy of consciousness some people deny altogether the importance of the human body as a condition of knowledge and other symbolic activities of man. To them body is just a dispensable adjunct to consciousness. The other negation goes to the other extreme in the name of scientific method. It is said that all human activities, including the "internal states" of the human mind, may be satisfactorily explained purely in terms of environmental conditions, organic reactions and "reinforcement".[12]

Body may be a dispensable adjunct to consciousness in a transcendental sense. But it is very difficult to understand the identity of a man without his body. What is transcendentally trivial or irrelevant, is for us absolutely necessary on the empirical plane and for all practical purposes. Man knows himself as a bodily being. Man *qua* man cannot even conceive his transcendental being. Without his body it is difficult for him to conceive or imagine his individual identity. A man without a body and therefore without his organs cannot possibly have symbolising ability to identify, rather to reidentify himself as a man. The concept of a bodyless man is an illegitimate abstraction from the experience of embodied man. This abstract concept has little relevance and no significance; for we cannot even imagine a single case of its application or instantiation.

The other extreme is equally untenable. The primacy of body does not entail denial of consciousness or self-consciousness. Apart from the well-known arguments of freedom, immediate experience and reflective memory, we can also plausibly use the argument of the continued existence of *other* human beings. Instead of raising the sceptical question, "What am I?", one might raise as well an equally pertinent question, "Who are they ?" Other human beings constitute a condition of my being human. In other words, this brings back the old question: whether the social dimension of the human life is a dispensable adjunct or an integral aspect? What enables a man to identify himself as a human being is in principle identical with what enables him to identify other human beings. Man is man-in-relation to *other* human beings. Abstracted from this relation what he is we cannot know. True, we can imagine that. But again that imagination would be an abstraction from or extrapolation of what man-in-relation-with-other-human-beings is.[13]

The social identity of man, rightly understood, explains human freedom. Neither the mental state is a *function* of the human body nor the human behaviour a function of genetic and environmental conditions. Man's response to the external stimuli brings about a significant *transformation* into the very nature and function of those stimuli. Man's *information* regarding the origin and process of those stimuli provides him an access to their nature and operation and enables him to influence thereby both. Society is a field of endlessly complex feedback operations.[14] The presence of the human phenomenon in the field introduces an irremediably "erratic" or subjective factor in the complex field which could be otherwise correctly comprehended and predicted by an all-knowing god. But, unfortunately for the radical behaviourist, the rules and regulations which constitute the social structure are themselves value-loaded or normative in character. Consequently, social determination itself proves to be an evaluative factor and it cannot be faithfully described by an external spectator. It is mainly for this reason that many thinkers are of the considered opinion that social phenomena have to be *understood* and cannot be *explained.*[15]

From Social Anthropology to Epistemology

The social identity of man has a very interesting bearing on the character and scope of the human knowledge. The matter may be looked at from two different but related aspects. Firstly, what man *is* certainly influences what man can possibly *know.* Secondly, the validity or otherwise of man's knowledge-claim of the world depends to a great extent on some socially acceptable (and therefore, questionable) rules and decisions. The second point may be elaborated into what is known as sociology of knowledge. The conditions of valid knowledge are undoubtedly psychological in "origin"; but the more interesting part of the process of establishing or dis-establishing the validity of knowledge-claim is of course social and objective. (For the time being I am dogmatically rejecting the view that knowledge is unconditionally valid.) The human body is not only a necessary condition of what he is, i.e. his identity, but also of what he can possibly know. Our mind is obliged to work within the boundary conditions set by our body. These conditions may be flexible but not boundless. This important factor always keeps the question of cognitive validity open.[16]

The categories and concepts used in organizing our experience are of dual character. From the standpoint of the individual man they should be taken as *a priori*. Yet, strictly speaking, they are not *a priori* in the Platonic or even in the Kantian sense. For their career and fortune are subject to the constraints of reality as communicated through the contents of experience. Man never encounters reality *qua* reality. His body, mind and society work as a sort of filter-cum-cushion in between him and the real world. Language is an integral part of man's social life. Categories and concepts are embedded in it. Almost every part of language has a bearing on the mode of man's experience of the world and also on his action vis-a-vis the latter.[17]

The conceptual framework of our knowledge and the reference framework of our action are all subject to historical wear and tear, addition and alteration. We cannot visualise an ideal framework either of thought or of action which will be able to adequately meet the demands of all actual thinkers and for all time to come. It is true that our natural language contains rules for the formation of meaningless or senseless sentences. But it is equally true that the artificial languages which try to rule out the possibility of formations of some such expressions are themselves not potent enough to generate the expressions of all types of our experiences, especially those which have value-load in them. Nor the so-called logically neat languages are adequate even for their own limited and proclaimed purpose.[18]

The quest for the hidden logical structure of natural language turns out to be elusive for several reasons. First, it tends to deny the ontological commitment of natural languages. Secondly, the attempt to deontologize or dilute the *referential* character of logical language proves abortive.[19] Finally, one is struck by the gradual discovery of a system of rules enabling one to penetrate the phonetic disguise of other man's thought. Successful actions and communications prompt us to believe that in terms of a language man can not only encode, but can also decode the intentions of his "ghostly" thoughts and "mysterious" encounters with the hidden reality. By virtue of his insight and symbolic ability man can penetrate the orthographic clothing of reality.[20]

Man in the Quest of Ought

The identity of man is not static. He wants to exceed himself. The self-exceeding enterprise of man has been explained in two different ways, *as a matter of course* and *as a matter of principle.* The history and civilization of mankind has often been interpreted as how the successive generations of human beings, using their knowledge and skill, science and technology, have understood and fought the forces of nature and established themselves in the position we find them today. There is another way of looking at this process. It would be claimed that a definite principle has been consciously or unconsciously motivating and sustaining human beings in their ceaseless efforts to remove the fetters, put up by the adverse forces of nature and society and to achieve more and more freedom. The latter interpretation does away with the distinction between *is* and *ought.*[21]

The first interpretation of history is open to the criticism that it negates human freedom. If the course of history is governed by certain *laws*, as distinguished from *principles*, and if the operation of those laws is independent of human action and valuation, it is not clear how man can be held responsible, praised or blamed, for what happens—good, bad or indifferent— in the process of history. History is inevitable.[22] But in that case human action and thought are irrelevant to its predetermined course. This reopens the age-old question of the relation between naturalistic (or antinaturalistic) determinism and human freedom, between what happens and what might have been prevented by timely human effort and initiative. Should we take whatever happens as inevitable? It is true that the theory of universal determinism can be reconciled with the freedom of *the finite* human being. Further, by affirming the limitation of human knowledge and/or defining the human freedom itself in terms of the recognition of necessity, it is always *possible* to vindicate the human freedom in *principle.* But the question is what is the character of this *possibility* and of this *principle*? Are they not purely metaphysical, in the bad sense? Can we indicate an *actual* human being who is really human and yet knows that he is *not* free? This freedom by definition rules out the existence of such a human being.

The self-exceeding character of man may be viewed in another perspective. I mean, the sociological perspective. While the historical perspective highlights the macro pictures of the achievements of mankind spread over several centuries and this obliges us thereby to underrate the significance of the individual human efforts and actions to fight against the forces of injustice and the enemies of freedom, the sociological perspective concentrates on the details of the human motivation and action, social interaction and process. Every approach has its own advantages and limitations. In the still framework of time, man's actions and efforts to mould his own destiny can be closely scrutinized and evaluated. Man needs a social structure to sustain him in all his actions. He finds himself within it and it is integral to him. He may like it or not. Nourished and sustained by it, generally speaking, he partly accepts it and rejects it partly. By merely accepting it he cannot ensure its perpetuation; nor by rejecting it he can destroy it.[23] But his acceptance and rejection have an impact a transformative impact on the social structure. His society is a constant object of his value judgement. He can always think of a fairer society, a more just society and/or a more free society, than he is living in. If he is a man endowed with leadership quality, his thought, translated into action, can bring about significant changes in the social structure, composition, and direction. Even if he is otherwise an ordinary man, his thought contributes to the creation of pro-change climate of opinion in the society. Thought is no less potent than action as a harbinger of the unborn society. In fact thought has a native tendency to transform itself into action.

Man in the Quest of Perfection

The self-exceeding character of man is evident not only in history and sociology but also in other modes of experience, e.g. religion and aesthetics. Thinkers have been persistently accused of romanticism and utopianism for their models or blueprints of the ideal society. It is interesting to note that Plato's *Republic* has been always criticized and yet again and again read and re-read. The concept of *Ramrājya*, to take another example, has been often characterized as utopian or unrealistic; even then in every age we have some people or other to sponsor and defend this ideal. Perhaps the most interesting example of man's incurable idealism is provided by the history of anarchism. Always dissatisfied with the state authority, the idealist man has never ceased to dream of an ideal society without it. The idea of the stateless association of free human beings, without let or hindrance from any external authority, has been always fascinating the finest of the human minds down the centuries.[24] Sometimes, this ideal has been portrayed within a religious framework and at times in a secular manner. We have been told of the City of God as well as of the stateless communist society. It is interesting to note the persistent fact that these ideals have proved to be elusive till date never been used as an argument against the importance of the concerned ideals as such. How are we to understand this apparently "anomalous" position? Man's worth is judged not necessarily in terms of what he achieves but perhaps also in terms of what he hopes.

The World-State has not been established; yet we are all enthusiastic about an international order. The League of Nations has died a premature death; yet we have the UNO. True, it is not in an ideal shape. But no right thinking man would dare call it useless and demand its dissolution. Kant dreamt of *Perpetual Peace.* However, in the last 200 years or so, we have had the experience of innumerable wars, local as well as global. Still we are working for disarmament and welcome the detente as a prelude to peace.

We want not only a peaceful world but also a beautiful world. We want that our lives, both individual and collective, should not only be perfect but also pleasant. We want peace and happiness. The haunting spectre of poverty and misery could not destroy our hope for better days.

Hoping is inherent in the human nature. Man knows that he is mortal. He lives his life under the shadow of death, yet he talks of immortality, develops elaborate systems and detailed arguments in support of his understandable belief in the immortality of soul. Man hates the *idea* of death. He dislikes it as it makes him deeply sad. However, it is also true that the shadow of death brightens up and deepens his consciousness and makes it more creative. It reveals one of the most noble and yet "unrealistic" dimensions of the human being.

III
MAN IN THE QUEST OF A JUST SOCIETY

The question of the human identity is fundamental in the matter of taking a correct decision on the nature of the ideal society, whatever that might be. Unless we have a rational and critical view about our own nature, it will be an idle exercise to speculate about the future of mankind, the elements of a just society and the other equally important and relevant issues. The true nature of our *existence,* rationally understood and critically evaluated, could give us an idea of what we can *possibly* achieve in a definite socio-historical context. One might question my attempt to relate our *existence,* what we are, to our *ideology,* what we can *possibly* achieve. I have already tried to answer this question when I have said that human existence is unique—unique in the sense that it has more potential than what is disclosed of it behaviourly and otherwise. The human existence is multi-dimensional, a bottomless pit or, to change the metaphor, an endless height. The limits of the metaphor have to be clearly borne in mind. We have also pointed out that man is finite, fallible and mortal. Here lies the ambiguity of the human identity. While his consciousness and action are functionally subject to many conditions, in principle his potentiality is infinite. And therefore we find him perpetually a self-exceeding being.

Man's fallibility, freedom and mortality rationally suggests what sort of a just society he should try to realize. Because of his mortality he would naturally like to achieve the aims and objectives which he himself believes in and shares with his fellow human beings. Time, especially his own life-time, is very important to him. This does not necessarily lead to a myopic and hedonic ideology; for his self-exceeding character is no less influential than the time which heavily weighs on his being. Secondly, the fallibility of man is the strongest available argument in support of a truly democratic society. If the philosopher might fail, there is no rational argument why the king could do no wrong. Constitutional insulation of the king or of the dictator is not a good enough premium to insure the future of democracy. Separation of power, a system of check and balance, and public accountability, irrespective of the formal education and wealth of the concerned public, are some of the basic elements of a just and truly democratic society. And, finally, this brings me to the most important element of the ideal society, i.e. freedom. If man, who is born free, is not allowed to question the rationale or otherwise of the fetters, found or used, to curtail his own and others' freedom, the very foundation of a just society is objectively attacked. Freedom is not to be defined or understood in terms of equality, still less in terms of equal opportunity of all human-beings, irrespective of their status, role and power. Though man is born rational and free, obviously not because of Aristotle and Rousseau, he cannot develop the former and enjoy the latter in isolation, disregarding what happens to the freedom and reason of his fellow human beings. In other words, man cannot conserve his most basic characteristics and values without objectively ensuring the same for the community or society of which he himself is an integral part. In view of the resistance put up by the reactionaries and other vested interests, how to ensure it practically? There is no readily available answer to this question which will be applicable to and acceptable for all human beings regardless of their social contexts. Only one thing perhaps could be safely affirmed, i.e. the method of trying to establish a just society must be rational and therefore questionable. But there is no guarantee that it will be necessarily peaceful.

REFERENCES

1. Julian Huxley, *The Uniqueness of Man,* Chatto & Windus, London, 1941. See also, Pierre Teilhard de Chardin, *The Phenomenon of Man,* Fontana Books, London, 1971.
2. D.P. Chattopadhyaya, *History, Society and Polity,* Macmillan, New Delhi, 1976.
3. Robert, E. M. Wheeler, *Archaeology from the Earth,* Clarendon, Oxford, 1954; see also, V. Gordon Childe, *What Happened in History,* Penguin, London, 1960.
4. Gordon R. Willey and Philip Phillips, *Method and Theory in American Archaeology*, University of California Press, Berkeley, 1958.
5. Frank Hole and Robert F. Heizer, *An Introduction to Prehistoric Archaeology*, Holt, New York. 1965.
6. Ernst Cassirer, *Philosophy of Symbolic Forms,* 3 Vols., New Haven, 1953, 1955, 1957.
7. Noam Chomsky, *For Reasons of State,* Fontana, London, 1973; see also, B.F. Skinner, *Verbal Behaviour,* New York, 1957.
8. See Chomsky [7] above, "Psychology and Ideology".

9. Harold Borko, ed., *Computer Applications in the Behavioral Sciences,* Prentice-Hall, Englewood Cliffs, N.J., 1962.
10. Noam Chomsky, *Current Issues in Linguistic Theory.* The Hague, Mouton, 1964; Roger Brown *et al., Psycholinguistics.* Free Press Paperback, NY., 1970; Pride and Holmes, ed., *Sociolinguistics,* Penguin, London, 1972.
11. Herbert Feigl, 'The "Mental" and the "Physical"', in Feigl *et al.*, ed., *Concepts, Theories, and the Mind-Body Problem, Minnesota Studies in the Philosophy of Sciences,* Vol II, Minnesota Press, Minneapolis, 1958.
12. See Skinner [7] above.
13. D. P. Chattopadhyaya, "Anthropological Rationalism" in *Individuals and Worlds*, Oxford University Press, New Delhi, 1976.
14. Heinz von Foerester, ed., *Cybernetics, Circular, Causal, and Feedback Mechanisms in Biological and Social Systems,* 5 vols., New York, 1950-55.
15. May Brodbeck ed., *Readings in the Philosophy of the Social Sciences,* Minnesota University, Collier-Macmillan, London, 1968 ; G . H. Von Wright, *Explanation and Understanding,* Cornell University Press, Ithaca, 1971.
16. Maurice Merleau-Ponty, *Phenomenology of Perception,* Routledge & Kegan Paul, London, 1970.
17. Sir Karl Popper and Sir John Eccles, "Falsifiability and Freedom" in *Reflexive Water: The Basic Concerns of Mankind,* ed. by Fons Elders, Candor Book, 1974.
18. Stephen Körner, *Categorial Frameworks,* Basil Blackwell, Oxford, 1974; Ernest Gellner, *Cause and Meaning in the Social Sciences,* Routledge & Kegan Paul, London, 1973.
19. Maurice Merleau-Ponty, *The Structure of Behavior,* Boston, 1963. Talcot Parsons, *The Social System,* Free Press Paperback, New York, 1964.
20. Jerrold J. Katz. *Linguistic Philosophy: The Underlying Reality of Language and its Philosophical Import,* George Allen & Unwin, London, 1971.
21. G. H. Von Wright, *Norm and Action,* Routledge & Kegan Paul, London, 1971
22. Isaiah Berlin, *Four Essays in Liberty,* Oxford Paperbacks, London 1969; E. H. Carr, *What is History,* Penguin, London, 1965.
23. D. P. Chattopadhyaya, *Individuals and Societies: A Methodological Inquiry,* Scientific Book Agency, 2nd ed, Calcutta, 1975.
24. John Passmore, *The Perfectibility of Man,* Duckworth, London, 1972; George Woodcock, *Anarchism: A History of Libertarian Ideas and Movements,* Meridian, New York.

2

ANTHROPOLOGICAL RATIONALISM : A REVIEW

I
"MAN" AND "WORLD"

Man lives in the world. The world environs man. Both *man* and *world* are ambiguous terms. Their clarifications are called for before we proceed further.

Man is embodied. The body of man is not homogeneous. It consists of different constituents and has different powers in it. First, certain constituents of the human body are natural and explainable in terms of the laws of nature. Secondly, some other constituents are biological and which without using some additional concepts and laws cannot be explained *adequately.* The naturalist explanation of the biological properties of the human body may be *clear* and even *interesting,* but not adequate. For certain properties of the human bodies remain unexplained or unclarified by the concepts and laws of the physico-chemical nature. Thirdly, apart from his bodily and the biological properties man has some psychological properties as well. At this level too a *reductive explanation* can be resorted to. But again that leaves certain psychological characteristics unexplained or half-explained. Fourthly, man, at times, is credited with parapsychological or paranormal characteristics and competence. The claim is that, apart from the empirically provable characteristics of man, there are reasons to believe that he has some other "subtle" and "higher" powers of knowledge and actions. Finally, man is said to have a social *orientation* not only in his psychology but also in his physiology. In other words, the body of man is sociologically *nurtured* and oriented. Culture has a say on the *human* body nurtured by it.[1]

Now in order to free the concept of world from its accompanying ambiguities, some remarks from the other end of the man-world relationship right to be looked into. First, world constitutes of *physical* and *chemical* properties. These properties are connectable and explainable by certain laws. And there is a persistent belief that these laws can be shown to be a consistent whole. Secondly, the physical and the chemical properties of nature through interaction and in the process of evolution give rise to a set of biological properties. At the biological level, the behaviour pattern of both plant life and the animal life, certain properties are discernible which are absent in the physico-chemical level. The abilities to move, interact, receive, preserve and use informations in a relatively organized and purposeful manner become evident at the biological level. Thirdly, the world we live in has certain *social* properties and functions which are more or less distinguishable from and yet not unrelated to the physico-chemical and the biological properties. The social life of man is in a very strong sense a part of the objectively given world. Society has an identifiable life of its own which is relatively independent of the lives of its individual members. Finally, we often speak of a realm of *values* and which constitute a part of the world. The values which man recognizes, follows or disputes are also identifiable independently of the lives of the individuals concerned.[2]

From one end one can say that man has physicochemical, biological, psychological, sociological, valuational properties in him, and, from another end, one can say that the world has "corres-ponding" physicochemical, biological, psychological, sociological and valuational properties or levels in it. These two ends are obviously related but their correspondence is neither necessary (i.e. invariable) nor unique (i.e. one-to-one). It is interesting to look into the presuppositions of this relationship. One of the main tasks of theoretical or philosophical anthropology is to do so.[3]

II
SOME IMPLICATIONS OF MAN-WORLD RELATIONSHIP

Such terms as man and world are obviously linguistic or symbolic devices invented and used by man to understand, among other things, the man-world relationship. In a sense, therefore, one might say, the man-world relationship, strictly speaking, cannot be understood in a language-neutral way. The non-availability of language-neutral account or image of the man-world relationship has some deeper implications.[4] First, man can well think of a manless world. In fact the pre-human, for example, the geological, past of the world is a scientific image of the world drawn by man himself. From the cosmological point of view too one can easily think of a manless world. True, in that case we are using the term "world" in the physicochemical sense. We can also think of the world as evolving from the pre-human level to the human one. Secondly, we can think of manless world in another way. Though man is in the world and as a part of it, but his presence may be ignored or remain unfelt or ineffective. As if man is dead in the world. The presence which is ignored, unfelt, or ineffective is not same as absence. It is somewhat like the presence of a *dead* body in a *living* room. This presence is *somehow* different from the presence of the furniture, curtains, and gadgets in a room. One can always say, "the world of dead man is not a manless world." Thirdly, there is another way in which one can think that the world is manless. Suppose man is there in the world but is so *absorbed* either in the processes of world itself or in his own, exclusively own, ideas and actions that nobody can feel that he is in the world. Man's absorption in the world or in himself may be so thorough that one can honestly and justifiably believe that the world is manless. In certain social situations when a man is expected to do something in order to make his presence somehow felt but remains *absolutely* indifferent and inactive, then people say, "there is none here" or "it is really a lifeless situation". Finally, man can think of a world from which man has totally *disappeared.* In science fictions, for example, we are often told of a world destroyed by a nuclear holocaust and from which the human species has disappeared. This is a possibility which is now so distinct that it is being seriously discussed by the people who are more attached to facts than fictions. Realists keep on warning us that this world can well survive a nuclear warfare but it would be then without the human species in it. Different pictures of the *manless world*, on final analysis, turn out to be *products of human thought.* When we say this, we are not highlighting the obvious psychology or subjective aspect of the human thinking. What we are trying to highlight is that all attempts of man to think of "manless world" or "world without man" are grounded in certain human or anthropological presuppositions.

And the rationality of the presuppositions is sought to be established along different lines and at least some of these deserve close scrutiny.

III
A CRITIQUE OF OPTIMIST THEORIES OF MAN-WORLD RELATIONSHIP

If it is assumed that man, though related to the world, has a cognitive competence that he cannot only know the world but also know that his knowledge is valid. This raises certain problems. Unless man is deemed to be very perfect, let us say, formed in the image of God, it is difficult to understand how can he possibly know the world so perfectly, i.e. validly. Behind the optimist theory of knowledge *of* the world there is a pre-critical assumption that, though situated in the world, man's knowledge of the world is not affected in the least by his situation. In other words man knows the world *freely.*[5] Freedom as such perhaps cannot explain the claim that man's knowledge of the world is perfect. In addition it is to be assumed that the freedom is exercisable without hindrance or constraint. Can man's knowledge of the world be free from the feed-effects of knowing and the conditions which make it actual?

Once it is accepted that man's knowledge of the world is perfect, naturally the question arises, 'how can it be so?". One well-known answer found in the history of ideas is: (a) because man is *perfect,* and (b) in exercise of his perfection has *created* this world. The search for the answer to the question, "how man has known *truly* the world?," hinges on the presupposition that man has indeed known the world truly or perfectly. The strategy of this optimist theory of knowledge is very clear. First you assume that the world is known perfectly and then you "show" that it has been made possible because of the perfect cognitive competence of man. It may be put in another way. The ground of the perfect knowledge of the world, let us call it science, lies in man's capacity to create the world *as we know it.* This view is historically associated with the name of Kant. But there are several other thinkers both in the West and the East who seem to hold similar views: that is, world is known perfectly because it is created by perfect man in exercise of his perfection (of perfect freedom to know).[6]

Clearly this view of the man-world relationship is very optimist, crediting man with the capacity of creating the world. But this optimist view seems to be inconsistent not only with the common sense view of the world but also with the scientific view of the same. Neither the man in the street nor the scientist, generally speaking, thinks that his view of the world is perfect. In the face of this questioning attitude, one of the following two lines of arguments can be adopted. One : man's knowledge of the world seems perfect because it is *limited* to the *surface structure* of the world and determined by available empirical informations. In other words the optimism associated with the first veiw is defended by the scope of the human knowledge of the world. One might say this is the standard phenomenalist version of science. The less we say about the world the more certain can we be about the (objects of the) world. To be most certain the information content of our cognition must be limited to the bare minimum. It is difficult (but not impossible) to be mistaken in my claim (for example) that the next raven I will see is black. By implication the phenomenalist gives up the claim of having known the world as such or as it is in itself in its deep structures. In this case scientific certainty is defended by moderating the claim of the scope of scientific Age. Ernst Mach[7] and Rudolf Catnap[8] in some of their writings argued in favour of such a view. Two: the alternative strategy of the optimist epistemology is that there is a perfect correspondence between the surface and the deep structures of man's knowledge and those of the world itself. Man's *sense-experience* can grasp certain *surface* properties of the world and man's *essential power,* i.e. *reason,* can grasp the essential structures of the world. This alternative version of the optimist epistemology can be further strengthened by claiming that the structure of man and that of the world are *identical at bottom.* Given this view, the question of manless world or, for that matter, the difference between the manless and the human world does not arise at all. According to this theory the non-existence of the world does not entail any loss of its *basic* contents because the same is *essentially* retained in the very nature, in the deep structure, of man. To put it from the other end, according to this view, the appearance of man in the world or appearance from it does not mean any *essential* difference in *basic* contents of the world. For man is in the world no matter whether it is manifest or not. This view has been expressed in different ways, both speculative and not so speculative. When I am talking of the second alternative strategy of the optimist epistemology I have primarily Hegel[9] and, in some respects, Chomsky[10] in my mind. The evolution or the emergence of man in the world has been deemed by the Hegelian to be inherent in the very nature of the world itself. The difference between man and the world is internal. It is a sort of identity in difference (or *svagatabheda* in the Vedantin language). This optimist and speculative view of the man-world relationship can be defined and represented in various other ways. But since this view is not very influential at the time we need not go into its details. But it will be noted that other less optimist and more influential views on the man-world relationship is significantly related to it both historically and conceptually.

One persistent objection that is raised against the above view is that man, himself determined by the world and situated in it, cannot possibly know it perfectly. The truth of man's knowledge of the world is bound to remain limited and open to questions. Even if it is assumed that the basic structure of the world and that of man's cognitive competence do *perfectly correspond,* the assumption by itself does not prove that man can, in effect, know the world in an unquestionable manner. Besides, it is very difficult to show the rationality of the basic assumption—the assumption of the correspondence between the structure of man's cognitive competence and that of the knowable world.

Perhaps in a weaker form and *apparently* withdrawing the assumption of correspondence, this view of the man-world relationship could be defended. It may be claimed that man's knowledge of himself and of the world grows, grasps more and more general properties, and goes deeper and deeper. The progress of scientific knowledge as evident from the history of sciences, it may be argued, suggests that man's mind is somehow endowed with the capacity to explore or discover the structures of the world which are objectively given and not created or constituted by his knowing mind. On the basis of increasing informations, logically and experimentally processed, man can make his knowledge more precise. *Precision*, *generality*, and *predictive powers* of human knowledge are not grounded in any *metaphysical* assumption.[11] In other words, these characteristics of the human knowledge do not presuppose any metaphysical realism. Once it is admittted that the human mind, either on its own or as the representative of God on the earth, cannot make the world possible, the world as known and knowable by the human species, it has to be further admitted that the world (whatever might be its surface and deep structures) is there independently of its being known. We are *ontologically* obliged to believe that the world increasingly and more and more accurately mapped by the scientists exist independently of its being known. And yet the changing characteristics of the scientific knowledge of the world—*precision*, *generality*, *predictive powers*, etc.—oblige us to admit that we can never be certain that the *known* structures of the world are in fact the real structures of the *world in itself.*

Each of the pessimist theories of the man-world relationship has its different forms. The strongest sceptical form of it confesses its inability to assert or deny logically anything about the man-world relationship. The main three steps of its argument may be stated thus. First, man can never be certain about his proclaimed knowledge of the world. For the sense-informations which form the basis of the claim of knowledge are themselves unreliable. Secondly, man cannot be certain about his own self and his cognitive competence. For every attempt to know the self or mind, which is supposed to enable him to know the world and his own knowledge of it, turns out to be elusive, if not abortive. And finally, given the first two steps of the argument, there is no logical means to be certain about the ways how man and the world are related. This theory has received different formulations in the history of Philosophy. In India some Buddhist thinkers like Dharmakiriti defended such a view with considerable ingenuity.[12] And in the west the name of Pyrrho is associated with such a view.[13] But it should be noted here that neither the Buddhist scepticism nor Pyrrhonism proposes to defend pessimism in *practice.* Sceptical pessimism is primarily a *theoretical* position. It highlights the hopeless inadequacy of our theoretical efforts and logical enterprises to ascertain the correct relation between man and the world. From this premise some sceptics draw the conclusion that it is only through practical encounters with the world that we can arrive at a better understanding of the man-world relationship. But there are some radical sceptics who would decline even to proffer this additional claim in support of practice. Both theory and practice, according to them, are equally questionable as a guide to thought and action.

Among the modern European thinkers, it is perhaps Hume who has argued most persuasively both in respect of elusiveness of the knowing self and the questionability of all the theoretical claims of science. He would not deny that man can and in fact does know. His scepticism is confined to the quality of the title of knowledge. He is prepared to accept the conclusion of scientific knowledge but his critique of induction is intended to show, among other things, the hopeless inadequacy of scientific methods for the purposes of systematization and prediction of the informations about the world. If we feel more certain about the informative content of our immediate experiences, feelings and impressions, it is only because of the thinness of the very limited scope of their content. Whenever, we propose to go beyond the immediately given experiences, Hume tries to show, how uncertain the passage turns out to be. Hume's scepticism does not entail any practical pessimism.[14] *Theoretical scepticism and practical optimism can well go together.* This is an upshot of Hume's critique of scientific method and practice of historical research. Hume himself was an eminent historian and for a time was known perhaps more for his historical works than as a philosopher. Without questioning his basic theoretical apparatuses, many of his followers in the nineteenth century developed theories of progress and radical forms of optimism.

It seems Hume and his followers were convinced by their studies in history that the *growing* character of knowledge, marked by progress and regress, and the rationalist's view regarding the finality of scientific knowledge do not go together. Descartes, Newton and Kant, the three most influential spokesmen of natural philosophy of the seventeenth and eighteenth centuries, had one thing in common among themselves, an unbounded faith in man's competence to achieve unquestionable knowledge.[15] Even if this competence fails at times to yield definitive results it should not be taken as an inherent blemish of the cognitive competence itself. If the human mind fails to know the world, the failure according to Descartes, is to be attributed to some will powers in the self, disturbing the lines of knowing or because of some external interfering power. In brief, left to itself, man's mind can know the world unfailingly, clearly and distinctly. However, it is not always left to itself in its cognitive enterprises.

Kant has also tried to develop a definitive theory regarding the man-world relationship.[16] He ascribes to the transcendental self the supreme cognitive competence, but, at the same time, adds that it cannot work on its own. In a way he also shares one part of the sceptic view, namely, that the self is elusive. Though the self itself is not empirically available, its contribution is available in our acquisition of scientific knowledge of the world. According to Kant, scientific knowledge is confined to the space-and-time-bound surface structure of the world and the surface structure itself is a construction of the cognitive structure of the human mind. Kant's strategy of proving the certainty of scientific knowledge of the world consists in demonstrating that the mind which makes knowledge possible also makes the world (the ordered objects of knowledge) possible. In brief, *knowing is making* (in so far as the world of science is concerned). However for Kant the limit of scientific world is not the limit of the world itself. The world of God, Self and other transcendental entities is not available to science. But denial of the same makes science impossible. Therefore, Kant is obliged to postulate the world beyond the limits of science, but about its structure he prefers to remain non-committal.

IV
THEORETICAL SCEPTICISM AND PRACTICAL OPTIMISM IN THE MAN-WORLD RELATIONSHIP

One can reject the main thesis of the strongest form of scepticism, namely, that it is *impossible* to know the knowing self, if any, and yet feel sympathetic towards the Humean view that the knowing self is *elusive.* To me the whole question, as formulated by the radical sceptic and the *practical* sceptic, appears somewhat misplaced and yet suggestive. "Misplaced" because the quest for certainty in respect of knowing self is uncalled for. "Suggestive" because the negative outcome of the research, as pointed out by different forms of scepticism, does not make knowledge itself impossible. To put it positively, one can claim that the sceptic's critique of knowledge (a) shows the possibility of knowledge without any definite commitment regarding the nature of competence of the knowing self, and (b) keeps the question of its *practical* utility open.

This positive outcome of the sceptic's critique has proved influential and in quite different directions. The two most important versions of the thesis "knowledge without foundation" are (i) no-ownership theory of knowledge, and (ii) knowledge without a subject. The former has been argued most forcefully by Gilbert Ryle in the recent past.[17] But ingredients of the view are there in the writings of Hume, Kant, Mill and Wittgenstein. The main point of the thesis which is of relevance to the anthropological rationalist is that knowing can well be logically construed as a kind of bodily activities. Knowing activities, like all other bodily activities and the rules used to explain them, are social and variable. The activities which can be brought under the label of *knowing* are not the same in all times and places. The other interpretation (ii) of knowledge without foundation may be ascribed to Karl Popper.[18] His view of *objective knowledge,* an entity in (what he calls) the world, wants to get rid of the question of the origin or foundation of knowledge. One of the interesting problems in the history of knowledge has been to recognize the distinction between *knowledge* proper and some such kindred concepts as *opinion, belief,* and *imagination,* and to define the relationship between them. Systematic and ingenious attempts have been made to define *opinion, belief,* and *imagination* to lift the concepts above their subjective/ psychological moorings and accord them an *objective* status. "Authoritative opinion", "faithful imagination", "true belief", "justifiable true belief", "undefeated justifiable true belief", etc. are among the notable definitions of knowledge. Attempts evident in these definitions to enumerate the necessary and sufficient conditions of knowledge have proved futile for the same reason.

Each one of them is said to be vitiated by psychologism. By highlighting the growing and historical characters of knowledge, Popper believes that he has found a non-psychologistic solution to the old problem. The thinkers who mainly provide the historical back-drop of Popper's theory of objective knowledge—Plato, Bolzano, and Frege—could not show us clearly how knowledge is related to opinion, belief, etc. One of the primary reasons, of their failure is to refuse to take a close look into the history (and sociology) of knowledge. Each of them takes a particular paradigm of knowledge and then tries to show how is it to be *distinguished* from (and not related to) opinion, belief, etc. In a way they fail to take note of the *practically* recognized relation between the changing (both respectable and degenerate) forms of knowledge. One feels that this is an abstract and transcendental approach to knowledge. It fails to recognize how human beings in different times and places as a matter of fact use the words "knowledge", "belief", etc., or their equivalents. The failure is sought to be justified by drawing a sharp line of distinction between (a) logical semantics and (a′) social semiotic, and (b) illustrations of concept and (b′) uses of words. Popper's approach is relatively, but to my mind, not adequately flexible. The reason why Popper can, while they could not, make different forms of knowledge (respectable as well as degenerate) citizens of the same world is historical (as distinguished from transcendental). Popper's view of knowledge has been influenced both by sociology and physiology and he finds no good reason to think that knowledge is something transcendental and ahistorical. Since I have discussed the question elsewhere, I do not propose to go into it here in details.

The main difficulty involved in defining the correct relation between man and the world seems to stem out from an abstract and primarily theoretical approach which fails to recognize adequately the *practical* day-to-day relations in which man enters into the world. Man and world are inseparably, dialectically, and, one might say, cybernetically related.[19] As I mentioned earlier, different dimensions of man's existence are grounded in and sustained by the world. The world itself enters into the life and existence of man in different ways, biologically, sociologically and in more refined cultural ways (valuationally, for example). Man, strictly speaking, cannot get out or stand outside the relational network which makes him a part of the world. Even the most abstract and subtle powers of human reason do not enable him to take a completely *detached* view either of the *world as such* or of *man as such*. The main point which the anthropological rationalist tries to highlight is that human reason, being embedded as it is in a given body and society, cannot get out of its attending conditions. The concept of *pure knowing self* and that of *the world-in-itself* are rationally constructed by human beings. This assertion should not be taken trivially. The point is that the *human* origin of these concepts must not be forgotten. The classical rationalist who believes in *infallible knowledge* and claims, in addition, that it is possible to have infallible knowledge of the world-in-itself, seems to have forgotten that his own being and all activities are involved in an objective historical process. Man and world are two changing variables in a dynamic equilibrium. By implication it is clear that the situation in which man knows the world and acts is itself an inseparable aspect of all his knowledge and action. This is partially anti-abstract and basically a pro-historical approach to the understanding of the *real* human situation. A comparable situation obtains there in the description of "quantum" situation. Terms like "observed object", "observed instrument", "experimental conditions" and "experimental results" are abstract and disjointed parts of a single overall "pattern". One has to recognize that these terms weave a *changing theoretical pattern* and that recognition is a condition of its *practical communication.*[20] It is only as an inseparable part of the world that man is obliged to act in and know the world. The other end of the same phenomenon may be put thus : the world cannot be understood as if man is not in it or not a part of it. One might say: this is *trivially* true, for it follows from the very definition of "understanding", which by its very nature is "human". My point is: this observation is not quite correct. Once we recall other available definitions of "human" framed in terms of a *transcendental witnessing self* which, though a necessary presupposition of understanding, is not a part of it, or those of "understanding" framed in the analogy of God's *perfect* understanding, the incorrectness of the objection becomes clear, i.e., "self" may be "superhuman" and "understanding" "divine".

The immortal knowing self of the Vedantin or of Kant is neither a necessarily part of the world we live in nor is necessarily in it. Analysing at least some definitions of "God's understanding", it may be shown that the world we live in is not a *necessary* part of that understanding: for it may be a mere non-existent possibility in it.

Anthropological rationalism, unlike divine rationalism of Leibniz and constitutive rationalism of Kant, emphasizes the point that man's abilities to know the world are variously limited. The human reason is neither (Leibnizian) sovereign nor (Kantian) autonomous but fallible.[21] Its limitations are multiple. First, it cannot get into "the deep structures" of the world, the existence of which it is *obliged to postulate:* otherwise the search for laws of science turns out to be *arbitrary* (act of choice), if not *irrational* (decision). Secondly, its power of systematization is "impaired" by conflicting evidences or informations made available by sense-experience. Thirdly, there is no *a priori* way of ascertaining the possible objects of experience structured in the depth or even on the surface of the world. Fourthly, the rules (or conditions) which define the realms or structures of possible objects are themselves not fixed for ever. The fact that objects of experience often prove *unruly,* i.e., baffle our hypothesis, suggests that neither ontologically nor methodologically human reason has any direct passage or access into the detailed composition of the world. Besides, the very presupposition that the world has its fixed composition (which can be explored by man) is naive and uncritical. The structured composition of the world is not like a closed book (with objects printed on its pages in different sizes of letters) which, if opened, would be the same (i.e., identically readable) to all readers. Objects of the world, one might say, are being *continuously* composed, decomposed, and recomposed.[22] As the process is continuous and only structurally identifiable and re-identifiable, we can cognitively map and re-map the world and programme our activities more or less successfully. Finally, the anthropological rationalist's world-mappings prove *corrigible* by, and in a way *continuous* with, the past and future experiences of his own and those of others.

Given the limits of our reason to know the world, we want to marshal evidences in support of our knowledge-claim of the world. Evidential support is not the only support that is sought after and used to defend the knowledge-claim. Sometimes analogies and even metaphors are of use in defending a hypothesis about the world or a part thereof. For knowledge, or what we call knowledge, is very much dependent upon the clarification of its relation with other accepted theories and laws. Analogies and metaphors also contribute to the understanding of the implied content of a theory. Evidence, argument, analogy, model and metaphor may in different ways perform one and the same job, namely, indicate the position of a *theory* on the map of our world-view. And they also help us in choosing a programme of *action* in a given situation (as understood by us). However, this is not to deny that analogy, model and metaphor do not perform other jobs.[23]

It must be mentioned here that of the available evidences relevant to a theory those which are of a negative nature are more important than which are of a positive nature.[24] To support the universal hypothesis '"all diamonds are hard" that is not of much consequence to produce several pieces of hard diamonds. From the point of view of testing the said hypothesis discovery of a single soft (i.e., non-hard) diamond is extremely important. Since it is not possible for man to know the absolute truth-claim of empirical hypothesis, it is advisable for him to look for the weakness, vulnerable spots, of his hypothesis. In other words, since every factual theory has its potential falsifiers, the facts which, if discovered, would overthrow or prove, at least partially, its falsity, it should be our endeavour to look for those falsifying facts rather than the (facile) supporting ones. This methodological prescription, now sharply formulated by Popper, was known and used at the common sense as well as the scientific levels by other people in the past.

V
FALLIBILISM IN THEORY AND PRACTICE

The relation between *fallibilism* and *falsificationism* is very interesting. Since man is fallible, always liable to commit mistakes, it is very rational on his part to try to find out where he has gone or is likely to go wrong.[25] His eagerness and attempt to prove that he is right is not of much help to him. What can possibly help him most is to find out where his theory and practice are likely to prove wrong. The critical search for potential falsifiers and possible criticism strengthen our theories and programmes of action.

However, the *parallelism between cognitive theories* and *action programmes breakdown at several points*. First, at the theoretical level we can afford to be patient and live with unsatisfactory theories. Even in the face of disturbing, if not contrary, evidences, we are obliged to retain the theory. Because no other alternative, a less "disturbed" alternative, is readily available with us. But in the case of action programmes, we are not always free to keep them suspended indefinitely. Even while the informations relevant to a particular action programme at our disposal are not good enough to justify its choice, we are practically obliged to act. In other words, even while informations do not warrant good decision or do warrant only bad decision, we have to decide and act. For example, if we are called upon to construct quickly a strategic bridge for military reasons, we may not be able (for time-constraints and other unavoidable reasons) to gather the informations necessary for awarding the contract to the best possible firm. Even then for compulsive practical "reasons" we are obliged to award the contract to the readily available company, compromising thereby the quality specifications which in ordinary circumstances are strictly insisted upon. In other words, in the sphere of practice, we are generally guided by certain considerations which cannot ordinarily be regarded as *reasons*. That on the basis of *inadequate information* and in the face of disturbing evidences we are *obliged to act*, shows one thing very clearly and, that is, notwithstanding the said grounds of scepticism, we act as if the available informations are *good enough reasons for actions;* and to wait or search for *better reasons* in this sort of cases would be *practically unwise*. In the case of competing and alternative scientific theories and called upon to choose one of them, we can wait and compare their relative strength and weakness. This shows: (i) scientific theories are open-textured and action programmes close-textured, and (ii) the logic of theoretical discourse cannot be extended, without significant modifications, to practical discourse.

Man is fallible both in theory-construction and in drawing up practical programmes. But while it is reasonable for him to carry on with an unsatisfactory theory, knowing fully well that it is unsatisfactory, it would be unreasonable on his part not to execute the action programmes simply because he knows that the informations on the basis of which the concerned programmes have been drawn up are inadequate or imprecise. Freedom to construct alternative *satisfactory* theories is much wider than that of constructing alternative *satisfactory* action programmes.

One can draw at least two lessons from the partial breakdown of parallelism between the logic of thought/theory and the logic of action/practice. First, if the thesis of human fallibilism is pushed to its logical extreme in practice, it would be difficult, if not impossible, to act in time. Most of our actions are necessitated by time and have to be completed within a definite period of *time*. Once this important temporal or historical dimension of time is adequately realized, we will be in a better position to appreciate why a theoretical sceptic like Hume is not sceptic in history. The point may be even strongly reformulated. *The logic of practice suggests a positive cure of the main problem of theoretical specticism.*[26] The other lesson drawn is this. When the adequacy requirements of rational action are defined in a very abstract and general manner (as, for example, Kant did in terms of universality and necessity), we find them hardly satisfiable. Therefore, a man who for pressing practical reasons is called upon to act urgently, finds no guidance for action from the Kantian theory of rational action. One might point out that the objective of Kant's theory of rational action is not to provide practical guide (as it is ordinarily understood) to a man engaged or about to be engaged in action. Even if this argument is accepted, the point remains that an abstract theory of action is of little help in real action situation. And this is mainly due to the pro-Kantian intention to develop logic of action in the model of logic of theory-construction, maintaining, to the extent it is possible, a thorough-going parallelism between the two, and partly ignoring the demands for convergence of and interaction between theory and practice.

VI
COGNITION, PRACTICAL EVALUATION, AND ANTHROPOLOGICAL RATIONALISM

The anthropological rationalist persuaded of the difference between cognitive enterprises and practical ones, is opposed to the idea of following the same logic in two different fields. Man, being what he is, i.e., fallible, is likely to go wrong both in knowledge and in action. In the case of knowledge he has the freedom to work patiently, taking his own time, but in the case of action often he has to decide and act quickly. True, man is engaged in search of *truth* in knowledge and wants to be correct or *successful* in his actions. Positively speaking, one might add, truth of knowledge and success of action are related both conceptually and practically. But none of these considerations warrant (a) subsumption of the logic of action under that of cognition, or (b) denial of the partial autonomy and the primacy of the logic of action. Man's ways of defining his relation with the world are diverse. In terms of theoretical awareness alone, i.e., disregarding the demands of practical awareness, no adequate definitions of the man-world relationship can be formulated. The fallible man cannot construct an infallible theory describing the natural world or a part thereof.[27] Nor can he ascertain *exactly* his relation with others in the world *exclusively* in the light of his theory. But, as I have tried to show, he must have theory to understand his correct position in the natural world, history and society. Without this theoretical map, however questionable it might be in principle, he does not have even a minimal guidance to act in the world. To discharge his practical obligations, a consequence of his being in the world, he must have a theoretical picture of his own self and also of the world in which he is situated. In fact the picture of his own self and that of the world he is in are two aspects of the same picture. As a theoretician or as a knower, if his picture is a little fuzzy, it does not disturb him very much. But for the purpose of living and acting *authentically,* he needs a clear picture which unfortunately is not available. For, besides other reasons, the co-painters of the picture, his fellow fallible human beings, are *objectively* responsible for it. If the social world is more fuzzy than the natural one, it is not because that our fellow human beings are non-cooperative by design. They themselves also experience a gap between the world as it is and the world, as understood and lived by them. That apart, the pictures of the world of our fellow human beings are not identical. This is evident not only from social sciences, in general, and anthropology, in particular, but also from natural sciences. It is perhaps relatively easy to correct the *human* pictures of the *natural* world. The theories of natural sciences give us and are tested by relatively *stable* informations of the structure of the world. In technical language, one might say, the potential falsifiers of scientific theories are definite, at least in the short-term. But when we come to the question of the programme of action in the social world, we find it difficult to remove its fuzziness or the indefinite character. In the social sphere the informations which can possibly improve the quality of our programmes of action are themselves relatively in flux.

In order to map our position in the social world and decide our action programme therein, we have to depend not merely on *cognition* but also, perhaps more so, on *evaluation.* In our theoretical zeal often we tend to forget the *gap between cognition and evaluation* and also the role that evaluation is called upon to play to narrow it down and reduce the threat of scepticism in the field of practice.[28] The evidences and the informations that we marshal and use to construct theory have no intrinsic measure or weight of their own. Evaluation is bound to be intuitive to a certain extent and particularly at the initial stage. The weightage attached to the information is partly determined by the intuitive evaluative activities of our mind. Our mind, sociologically influenced as it is, cannot evaluate the informations in a culture-neutral manner. The true image of man cannot be drawn in an abstract manner, disregarding his biological roots and cultural affiliation. As an animal, man has to adapt himself continuously to his changing and at times challenging environment. Though culture through its different standards and institutions, ways of punishment and reward, have helped man to decide his attitude to different practical as well as theoretical issues, yet he has to rely considerably on his intuitive and non-measurable evaluative resources.[29] In between Yes (or assent) and No (or dissent) the spectrum is quite wide and one's evaluation has an important say in determining the placement of one's decision or conclusion within the range of that spectrum. Given the same informations, different men, in all probability, would more or less differ in evaluating (the content of) informations and in the details of the operative part of the resulting decisions. The fact that, in spite of their *individual* difference in evaluation of informations, they can and do live and act together in the *community* sphere, is full of significance. And this interesting fact is to be understood in terms of the needs and influence of customs, conventions and social movements of the concerned community life or culture. It is indeed an instructive phenomenon that *differently informed* people *practically* live the *same* community life. This is true more or less, both of laymen's community as well as professionals' (scientists' and philosophers') e.g., community. The anthropological rationalist takes pains to show how man being aware as he is of the limits of his reason, tries to learn from experience and mistakes, both of his own and also of others, tolerates, and lives with difference both in theoretical persuasion and practical pursuit. This seems to me a *true description of the human situation.* Portraying man as more rational than what he really is or than what is evident from the history of his ideas and actions, we may glorify and flatter ourselves. But that does not improve either our theoretical competence or our practical abilities. On the contrary, this false picture of man morally oppresses and bores us, generates false and unsustainable enthusiasm and optimism in us, and, in the process, impairs the quality of our efforts to know the world and shape it appropriately. The anthropological rationalist aims at giving a true picture of the man-world relationship and, at the same time, *improving* it.[30]

REFERENCES

I. The human body is a sort of living and moving hyphen between the natural world and the cultural one. The crisis of the modern natural sciences and that of the social ones cannot be properly understood without a clear understanding of the ways how the human body is contributing to and being influenced by the overall crisis. Besides Freud and Husserl, who have studied the problem in depth are Maurice Merleaup-Ponty *(Phenomenology of Perception,* tr. Colin Smith, Routledge & Kegan Paul, London, 1970, pp. 67-199; *The Primacy of Perception,* ed. James M. Edie, Northwestern University Press, Chicago, 1964. pp. 43-95) and Paul Ricoeur, *Freedom and Nature,* tr. Erazim V. Kohak, Northwestern University Press, Chicago, 1966.

2. Theodosius Dobzhansky, *Mankind Evolving,* Yale University Press, New Haven, 1976; D.P. Chattopadhyaya, *Individuals and Societies: A Methodological Inquiry,* Scientific Book Agency, Calcutta, 1973.
3. J. Bronowski, *Science and Human Values,* Messner, New York, 1956; *Sense of the Future,* MIT Press, Camb., Mass., 1977.
4. P. Feyerabend, *Against Method,* New Left Book, London, 1975; Thomas S. Kuhn, *The Structure of Science,* Chicago University Press, Chicago, 1962; David Bohm, *Causality and Chance in Modern Physics,* Routledge & Kegan Paul, London, 1957. The limits of reduction, translation, and mensuration, to my mind, deserve very careful analysis. The interesting point is that, in spite

of the limits, inter-disciplinary translation and inter-cultural communication are possible.

5. This view is to be found in the *Vedanta-Sutras of Sankara* (tr. George Thibaut, Motilal Banarasidass, Delhi, 1968). Elements of this view are found in the writings of Plato, Spinoza, Leibniz, and Kant. K.C. Bhattacharya's *The Subject as Freedom* is an ingenious defence of this view.
6. Besides the thinkers mentioned in (5) above, one can possibly refer to Ramanuja and Sri Aurobindo in this connection. The modal logicians and essentialist like Saul Kripke have been arguing substantially to this effect. See, for example, Kripke's paper "Identity and Necessity" in Stephen P. Schwartz *(ed), Naming, Necessity, and Natural Kinds,* Cornell University Press, lthaca, pp. 66-101.
7. Ernst Mach, *Knowledge and Error,* intr. Erwin N. Hiebert, D. Reidel Publishing Company, Dordrecht Holland, 1976. See, in particular, ch. XVI. "Presuppositions of Enquiry" and ch. XVII, "Pathways of Enquiry."
8. Rudolf Carnap, *The Logical Structure of the World,* tr. Rolf A. George, Routledge & Kegan Paul, London, 1967. See, in particular, Parts IV & V.
9. *Hegel's Logic,* tr. William Wallace and foreword J.N. Findlay, Oxford University Press, London, 1975, ch. V.
10. Noam Chomsky, *Problems of Knowledge and Freedom,* Fontana-Collins, London, 1972.
11. Karl Popper, *Conjectures and Refutations,* 2nd ed., Basic Books, New York, 1965; S. Körner, *Experience and Theory,* Humanities Press. New York, 1966.
12. F. Th. Stcherbatsky, *Buddhist Logic,* 2 vols., Dover, New York, 1962; K.N. Iaytilleke, *Early Buddhist Theory of Knowledge,* Allen & Unwin, London, 1963.
13. Norman MacColl, *The Greek Sceptics from Pyrrho to Sextus,* London and Cambridge, 1869; Mary Mills, *The Greek Sceptics, New York* and London, 1929.
14. T.E. Jessop, "Some Misunderstandings of Hume"; and Richard H. Popkin, David Hume : His Pyrrhonism and His Critique of Pyrrhonism"; in V.C. Chappell (ed.), *Hume,* Macmillan, London, 1968.
15. Gerd Buchdahl, *Metaphysics and the Philosophy of Science,* Basil Blackwell, Oxford, 1969. See, for example, Herman Helmholtz, *Popular Lectures on Scientific Subjects,* Dover, New York, 1952.
16. D.P. Chattopadhyaya, "Copernicus Betrayed" in *Journal of Philosophy and Research,* Vol.XXXII. No. 1, September, 1971.
17. Gilbert Ryle, *The Concept of Mind.* Penguin, London, 1963.
18. Karl Popper, *Objective Knowledge*; Oxford, 1972; Karl Popper and John Eccles, *The Brain of the Self,* Springer, New York. 1977; D.P. Chattopadhyaya, "Bolzano and Frege. A Note on Ontology" in *Logic, Ontology and Action* edited by D. P. Chattopadhyaya and P. K. Sen, Macmillan, New Delhi. 1979.
19. D.P. Chattopadhyaya, *Individuals and Societies : A Methodological Inquiry,* 2nd ed. Scientific Book Agency, Calcutta. 1973: Norbert Wiener, *Cybernetics,* 2nd rev. edn., 1962; Martin Shubik, *Games for Society, Business and War,* Elsevier, New York. 1975.

20. David Bohm, "Science as Perception—Communication" in Frederick Suppe (ed.), *Structure of Scientific Theories.* University of Illinois Press. 2nd. rev. ed , 1977, pp. 374-91.
21. D.P. Chattopadhyaya, "Reason: Sovereign. Autonomous. and Human" in *Individuals and Worlds: Essays in Anthropological Rationalism,* Oxford University Press, New Delhi, 1976. See also, D.P. Chattopadhyaya, *Knowledge, Freedom and Language,* Motilal Banarsidass, Delhi, 1989, ch. 4, pp. 34-58.
22. Objects are not permanent forms or frames of micro entities or, what one might call, subsensibilia. I am opposed to Whitehead's *pro-Platonic* theory of objects and inclined to accept the *process* concept of objects as defended by some quantum physicists-philosophers. Alfred North Whitehead, *Adventure of Ideas*, Free Press, New York, 1967, ch. XI, "Subjects and Objects"; W.V. Quine, "Referring to Objects" in *The Roots of Reference,* Open Court, La Salle, Illinois. 1973.
23. D.P. Chattopadhyaya, "Models and Metaphors in Arts, Science and Mathematics" in *Mind, Language and Necessity,* edited by D.P. Chattopadhyaya and P.K. Sen, Macmillan, New Delhi. 1981.
24. Karl Popper, *The Logic of Scientific Discovery,* Hutchinson, 1959, pp. 387-419; W.V. Quine "On Popper's Negative Methodology", in P.A. Schilpp (ed.), *The Philosophy of Karl Popper.* The Library of Living Philosophers. Open Court, La Salle, Illinois. 1974, pp. 118-20.
25 C.S. Peirce, "The Scientific Attitude and Fallibilism" in Justus Buchler (ed.), *Philosophical Writings of Peirce,* Dover, New York, 1955. For Popper's concept of fallibilism see "Truth, Rationality and the Growth of Science" in *Conjectures and Refutations,* London. 1969. See also Paul Ricoeur's *Fallible Man,* Regnery, Chicago, 1965.
26. See, for example, Hilary Putnam's "Degree of Confirmation" and "Inductive Logic" in *Mathematics, Matter and Method,* Cambridge University Press, 1975; Jürgen Habermas, *Theory and Practice,* Heinemann, London, 1974. Both Putnam and Habermas emphasize from different standpoints, the primacy of practice.
27. *The New Science of Giambattista Vico,* tr. T.G. Bergin and M.S. Fisch, Cornell University Press, 1970. Kant's theory of knowledge and especially the weaker forms of the Kantian view have been deeply influenced by Vico.
28. C.I. Lewis, *An Analysis of Knowledge and Valuation,* La Salle, Illinois, 1946. Issac Levi, *The Enterprises of Knowledge,* MIT Press, Mass , 1980,
29. Algorithmic measure of verification, falsification. and confirmation is not available. It is said to be impossible to devise a precise and complete step-by-step computational procedure expressible in a programming language.
30. Being a part of the world he lives in, man feels obliged, both naturally and culturally, to change it. In this respect the anthropological rationalist accepts Marx's view that the philosopher or for that matter every theoretician has a practical responsibility of trying to change the world.

3

Sri Aurobindo's Concept of Man

Man is a Ambiguous Phenomenon

Man can neither altogether reject his infra-mental past nor can he be completely aware of his supra-mental future. Caught in between the forces of Nature and the Spirit, man is evolving, growing and moving towards a definite goal. Man is a super-normal phenomenon in the normalcy of Nature, a mental being placed in, purposively related to, and influenced but not completely guided by Nature. Nature poses a sort of spiritual challenge to man and does it with a purpose, which she only partly reveals and mainly conceals, and of which man is peculiarly equipped to be conscious of on reflection.

Man is a thinking instrument of Nature. To put it in other words. Nature becomes self-conscious in Man. At the level of Matter, consciousness is in the form of nescience, or thick darkness, at the level of life it is in the form of ignorance, or melting darkness, and at the human level it assumes the form of Reason, a sort of twilight area. The ambiguity of human nature is accompanied by the twilight faculty of mental reason. The bright light of the Spirit is yet an unachieved ideal for man which leads him upward to the supra-mental plane of existence and consciousness. Man, the mental being, is a complex of different kinds of knowledge and power. "In mind itself there are grades of series and each grade is again a series in itself; there are successive elevations which we may conveniently call planes and sub-planes of the mental consciousness and the mental being."[1]

The Dimensions of the Human Being : Physical, Vital and Rational

True, all men are not equally developed, but humanity as a whole, says Sri Aurobindo, has arrived at no higher plane than the physical-mental, the lowest sub-plane of the intelligence. Generally speaking, we take our stand on which particular plane or sub-plane depends on the lower stages, and occasionally responds to the light from above. Usually we try to adjust ourselves to the requirements and forces of Nature and that explains why conformism and positivism are proving so popular and effective. From this it would be rash to conclude that there are no exceptional and creative human beings among us. Nature has her own plan in shaping the destiny of mankind, and sustains her initiative through relatively more gifted individuals among us. We are mutely desired by Nature to help her in her endeavour to realize the higher planes of the Spirit. The physical man proves of little help to Nature in her work to take us upward to the planes of overmind and supermind. For the physical man works with the physical brain, the physical sense mind and the physical sense organs which are said to be incapable of going deep into the secrets of Nature and can give us only, what Sri Aurobindo calls, a physical sense of reality.

The physical man has a vital plane in him which is dominantly instinctive and impulsive. These impulses and instincts are hereditary and customary, forming no disciplined order. Sensations, desires, hopes, feelings and satisfactions are all impulsive formations of life-consciousness, stimulated by and dependent on external things and contacts. These are all unmistakable characteristics of the vital man who is engrossed in the practical, the immediately realizable and possible, the habitual, the common and average mode of life.

The physical man has also his mental part concerned as it is mainly with the physical nature and its images. Objective, customary, practical and traditional are his ideas and activities. The superior values of life are, from this point of view, superfluous or at best helpful adjuncts. Strongly influenced and dominated by an immediate sense of utility or pragmatism, the mind of the physical man is more or less indifferent to the higher modes of thought and action and unresponsive to the finer activities and movements of imagination and feeling. To him whatever is not immediately useful is either a pleasant luxury or an abstract fancy.

Whatever, falls below or passes over the physical mind is regarded by the physical man as dispensable addenda to his life. Unaware of the deeper unity and continuity of the different natural planes and sub-planes and yet called upon by practical reason to account for them, often he resorts either to physical laws or to traditions and customs. He always tries to explain the unknown in terms of the known, and in the process of doing so commits the fallacy of *reductio ad absurdsum.* Unfortunately, this mode of explanation is often uncritically represented and/or accepted as the scientific mode of explanation. We are advised by Sri Aurobindo to remember that the laws and possibilities of physical Nature cannot be entirely known unless we know the laws and possibilities of the supra-physical Nature. A critic of the 'lower' scientific outlook, Sri Aurobindo does not deny the provisional necessity of this outlook as a starting point. But he believes that, sobered by its experience, science in its more developed form would leave behind its positive and pragmatic craze and be influenced by systematic reasoning and philosophical speculation which may not be of any immediate use and yet would successfully anticipate many future observations and experiments.[2]

Above the physical mind and deeper within the physical sensation, there is, what Sri Aurobindo calls, an intelligence of the life-mind. This is said to be more open, more than the physical mind, to the psyche (*cittapuruṣa*). It is an obscure life-soul, a frontal formation of the vital being (*prāṇa-puruṣa*) and not the psychic being as such. The vital man is concerned mainly with the life-being and the life-force. To him the world is a field for the fulfilment of life-power, impulsion, ambition and adventure. Physical existence is an inert form of the life-force and its significance is only instrumental to the expression and satisfaction of Life which is based on the 'secret subliminal vital being'. The vital man may realize that subliminal vitality which considerably helps the working process of evolutionary Nature. Partly free from the bondage of the blind physical universe, he can understand the inner Life of Nature within and without him, 'independent of the body and of the symbols of the physical world'. The vital man betrays two very important characteristics, passion and the instinct of possession. The vital man, the *rajasika puruṣa*, is interested mainly in a life of action and not in one of contemplation.

As the physical plane of human evolution has sub-planes in it, the vital plane has also sub-planes within its fold. The first subplane is 'a dumb inconscient drive or urge', an imprisoned will, an extremely rudimentary form of consciousness which is unconscious of its own consequence and prospect and completely under the influence of the Universal Will. This sub-plane contains the seed of individuality in an obscure and unformed state. The second subplane may be characterized as the ground of desire. It is at this plane that the completely imprisoned will of the previous subplane partly raises its head, becomes partly liberated, but being limited as it is in its capacity cannot possess all that it is eager to be. In the third sub-plane the eagerness to possess ceases to be blind and dumb. Being partly conscious of its potentiality and prospect, Life now knows that to get more it has to forget something which is immediate and grossly sensible and accordingly becomes prepared to sacrifice the present for the future. It is out of this knowledge and preparedness that a new power emerges. This is the power of Love: 'The-bud of the third (vital sub-plane) is Love', as Sri Aurobindo puts it. It is in and by love that life goes beyond itself, comes in communion with other forms of life and thus becomes more adequately conscious of itself. In the fourth plane, the bud of love blossoms into a flower of perfection. Here the original will of Life emerges in its fullness and with purity. At this plane the aspirations and desires of life are truly illuminated and fulfilled.

Whatever might be the graded achievements of Nature in the plane of Life, it is not quite aware of the light and potentiality which it carries within itself. The object of life is not simply living: it has to outlive what it immediately finds around itself. Paradoxically enough, to be full and perfect, Life must go beyond itself. Matter for its, own apparent-inherent inconscience and Life for its own apparent inherent ignorance confuse their autonomy with the sovereignty which belongs only to the Spirit. As the potentialities of Matter cannot be realized with the materials readily found within itself, the potentialities of Life cannot be fully realized with the means and mechanism supplied by it in its pre-reflective stage. It is to be remembered in this connection that every plane of Nature has a double purpose to realize. First, it realizes its own potentialities together with those of Nature as a whole. But in its pre-reflective stages Nature is not conscious of its simultaneously operative double purpose. Strictly speaking, no plane has its own exclusive potentialities. The actualities of one plane are correlated to the potentialities of another and the converse. It is only from the Spirit's point of view that all potentialities are truly his own. The process of their actualization is spread over all the conceivable planes and sub-planes of evolution. Sri Aurobindo points out that Man, remaining as he is almost exclusively confined within a less dark veil, is under the wrong impression that the different planes of Nature are completely independent and secular. Like every plane of existence, Man must exceed his individual bounds to realize his great inward potentialities. Man must affirm and exceed himself to realize what he really and essentially is. The Universal Will is engaged in preparing the grounds for man's ultimate self-affirmation. And that constitutes the rationale of the whole evolution of Nature—sub-human, human and super-human.

Above the level of vital mentality there is 'a mid-plane of pure thought and intelligence to which things are the most important realities'. The philosopher, scientist, intellectual, idealist and dreamer are at the top of this mid-plane. They are not necessarily fascinated, captivated and misled by the 'thingish' character of things and have the power to be aware of their potent and subtle ingoings. In the man of pure thought and intelligence, Nature is partly conscious of her own goal—the goal of which she is more or less unconscious at the sub-human levels. In these subhuman levels, there is no doubt physical and vital struggle and strife, but it is only with mental conflict that Nature inaugurates the human stage of her evolution. Man is the thinking instrument of Nature, relatively free from the rigid grip of vital laws, free enough to bring about an adjustment in those laws and to control them in a way that would be helpful for him in his quest for the highest ideal. If we remember the double purpose of Nature's evolution, we must admit that Man's ideal is at the same time also Nature's. That explains why man is not completely free to impose his own ideal on Nature's working process. Man may control the conflicting tendencies of his own nature but cannot totally transform their inner drift. Nature has not yet given a free hand to mental man who is still partly under the laws of Matter and Life. Though not so obstinately and insistently as Matter and Life, yet Mind in Man, under the influence of the first two planes of Nature, wrongly supposes itself exclusive and independent, and on the basis of this wrong supposition tries to realize its (great) ideal within the (narrow) limits of its physical-vital existence. Interested in evolving the higher planes of existence, including the Supermind, Nature cannot remain long an indifferent spectator and a silent witness to the defective and myopic endeavour of mental Man. Man is under an intrinsic, i.e., natural obligation to discover the inner harmony underlying the struggle and strife of the material and the vital planes of existence, and to shape his own life in the light of that harmony. If man becomes exclusively concerned with his own mental or rational conflicts and fails to see the light of harmony and to be guided by it, Nature intervenes. The sovereign rules of Nature ultimately over-rules the autonomous rules of Matter, Life and Mind.

Man is the middle-term of Nature's evolution. The more he realizes his own ideal, the more he upgrades his own status. But he has not yet discovered his own natural truth. He is still an abnormal in the normalcy of Nature, an imperfect apostle of a perfect ideal. But 'this imperfection', to use Sri Aurobindo's own words, 'is not a thing to be at all deplored, but rather a privilege and a promise, for it opens out to us an immense vista of self-development and self-exceeding'.[3] The process of self-development and self exceeding is tortuous and full of difficulties. The blind and dumb, obstinate and self-affirming laws of Matter and Life are not ready to open themselves to the twilight of human reason. On the contrary, they materialize and vitalize, i.e., try to downgrade and weaken the influence of human reason. Matter is comparatively simple in its organization and therefore it is with less difficulty that Reason can tackle it. But the opposition that Reason faces at the level of and from Life is rather stiff and persistent. In the struggle between Matter and Life on the one hand, and Mind, as we find in the mental rational being, on the other, the former slowly but inevitably lose their ground in favour of the latter. Of this silent struggle and its results, man may be (but is not necessarily) conscious. The more he becomes conscious of his higher purpose and opens himself up to the light above, the faster becomes the pace of his progressive march towards the goal of spiritual evolution. As every plane of existence is eager to maintain its status quo, Man's spiritual journey cannot be linear and easy. A proper understanding of this non-linear movement of Nature demands of us a thorough probe. The dynamics and rationale of this movement provide the spiritual silken thread connecting the different planes of existence cutting across their sensible boundaries and leading them to their common spiritual goal.

A Critique of the Concept : Scientific and Mataphysical

Sri Aurobindo's concept of man, particularly his account of man's position in the scheme of Nature, may be criticized on various grounds. In fact a charge of determinism has been levelled against his views. Sri Aurobindo himself is aware of the fact that this theory may be challenged on the ground that it does not have sufficient foundations or that it is highly speculative or metaphysical, and that its account of different planes of existence is superfluous. One may question the very usefulness of the idea of purposive evolution, and point out further that the ideal of human unity that Sri Aurobindo speaks of may be defended on a different ground—different from the one offered by him. Sri Aurobindo admits that the fact of the past and present regarding human and sub-human grades may be interpreted without making use of the concepts of teleology and evolution. Sri Aurobindo does not deny the relevance of the critic's point. In self-defence, he replies that his account of man in the process of evolution is based on true and infallible knowledge and facts of reality and that there is nothing theoretical about it. Theory being a product of human reason and often based on inadequate acquaintance with reality may be questioned, corrected and even refuted, but fact is either to be recognized or ignored and cannot be manipulated according to our sweet will. The matrix of all our doubts, questions and confusions regarding the bona fides of the evolutionary and teleological interpretation of history and pre-history of man is our relative ignorance, our lack of intuitive knowledge.

Sri Aurobindo is prepared to admit that even those who share all his fundamental views may refuse to accept his conclusion, namely, that there is a teleological purpose in creation. One is quite within one's right to argue that 'The Divine has nothing that He needs to gain or that He has not,' and that 'if there is creation and manifestation, it is for the delight of creation, of manifestation, not for any purpose.'[4] The teleological view of evolution has been sought to be countered also by the materialist. The materialist assumes that the whole evolutionary process is the work of an inconscient energy which acts automatically and mechanically and cannot have any conscious purpose in it.

While Sri Aurobindo recognizes the 'considerable cogency and importance' of the scientific or materialist objection, he clearly asserts that this 'view of things is not complete and its cogency is not conclusive'. He is not prepared to accept the Darwinian thesis that the evolution of man may be satisfactorily accounted for in terms of a series of variations. The evolution of species or forms does not shed light on the very necessity of evolution. Evolution may be better explained if the cause of variation can be ascertained, for it is the cumulative result of variation which accounts for the evolution of species. Darwin's account does not enlighten us on the matter. Darwin frankly admits his ignorance about the cause of variation. If he is pressed further he ascribes it to chance. As against Darwin, Sri Aurobindo argues :

> Even in the inconscient, there seems to be at least an urge of inherent necessity producing the evolution of forms and in the forms of developing consciousness, and it may well be held that this urge is the evolutionary will of a secret conscious Being and its push of progressive manifestation, the evidence of an innate intention in the evolution. This is the teleological element and it is not irrational to admit it; for the conscious or even the inconscient nisus arises from a truth of Conscious Being that has become dynamic and set out to fulfil in an automatic process of material nature...

Sri Aurobindo finds that it is relatively more difficult to meet the metaphysical objection for it is 'more serious' than the scientific objection. Logically it is undeniable that the Absolute, which is complete and perfect 'by definition', can have no purpose to realize in and through the process of evolution. And in that case, evolution turns out to be a mere expression of Divine Delight and absolutely objectless. Though this is a serious objection, Sri Aurobindo refuses to accept it. His first point is that material world is not absolute. Secondly, it is not complete. And, therefore, it is quite conceivable that evolution of the material world has something to gain through the process of evolution. Besides, when Sri Aurobindo speaks of purposive evolution, he does not use the term "purpose" in the ordinary human sense. The purpose that guides the process of evolution is 'the urge of an intrinsic Truth necessity conscious in the will of the indwelling Spirit. Against the objection that all this evolution is a game or *līlā* and therefore objectless, Sri Aurobindo argues that even a game has an object or purpose and without the fulfilment of that object it lacks in the completeness of significance.

> ...The material world is not an integral totality, it is a part of a whole, a grade in a gradation; it may admit in it, therefore, not only the presence of undeveloped immaterial principles or powers belonging to the whole that are involved within its matter, but also a descent into it of the same powers from the higher gradations of the system to deliver their kindred movements here from the strictness of a material limitation.[5]

Disposing of the more serious metaphysical objection against purposive evolution, when Sri Aurobindo turns to the scientific objections he finds that it is concerned only with the outward and visible machinery and process, with the detail of Nature's execution, with the physical development of things in Matter and the law of development of life; and Mind in Matter. However important the machinery or the exact genealogy or chronological succession of time may be, it is after all secondary—secondary to the plan or purpose which lends significance and direction to the process of evolution.

Sri Aurobindo's position on evolution clashes with that of Darwin and Lamarck. While Darwin thinks that the cause of variation is mechanical, Lamarck thinks that it is due to the conscious effort of the individual being to adapt himself to his environment. Darwin attaches supreme importance to the almost blind operation of mechanical laws. Lamarck believes that the cumulative results of the individual's conscious efforts are transmitted through heredity. Modifying Lamarck's position, some of his followers have argued later on that the change in the individual is effected not so much by his conscious efforts as by his unconscious response to the environment and the laws under the influence of which the environment itself changes.

The teleological account of the evolution has been rejected not only by Darwinians and Lamarckians but also by Marxists. On the nature and evolution of Life, the Marxist's position is a compromise between Vitalism and Mechanism. In common with the Vitalist, the Marxist rejects the Mechanist's view that life may be deduced from mechanical and chemical laws. He also rejects the Mechanist's view that the origin of life is due to chance. The Marxist accepts at least a part of the Mechanist's thesis when he denies the existence and operation of any ideal principle (like 'entelechy', 'soul', and 'life-force') in living creatures and organisms. He thinks that life represents something quite different from, and higher than, inorganic matter. He draws a line of qualitative distinction between 'material form of motion', and 'biological form of motion'. In this connection he refers to Engel's notion of Life: Life is the mode of existence of albuminose bodies and this mode of existence essentially consists in the constant self-renewal of the chemical constituents of these bodies.[6]

The Marxists share the Darwinian denouncement of teleology. But they themselves differ on a very important point. While the Darwinian lumps 'natural selection' and 'the survival of the fittest' together, the Marxist thinks that they are entirely separate. The latter argues, to illustrate his point, that selection by the pressure of over-population, where perhaps the strongest survives at the initial stage, may result in weakness of the survivors in many respects. Secondly, selection by greater capacity of adaptation to altered circumstances may well mark a sort of regress or degeneration of the survivors in some respects. At times we survive by downgrading ourselves.

Sri Aurobindo rejects the materialist accounts of the evolution of Man offered by the Darwinian and the Marxist mainly on the alleged ground that they do not adequately recognize the importance of human phenomenon in the scheme of the Universe. Apart from his general objection that Darwin and Marx have failed to see the deeper significance of teleology, he points out that their main principle of explanation, matter-in-motion, itself is not self-explained and therefore begs explanation. Matter itself can no longer be explained by Matter alone, for it does not appear to be self-explained.[7] The materialist's substitute for teleology is chance. If teleology does not suit his reason, one wonders how chance or accident can be accepted by him as reasonable and satisfactory. Teleology explains evolution from within the process as its internal dynamism; but the materialist theory approaches evolution from without and therefore goes deep into the process and accounts for its rationale. Sri Aurobindo does not deny the limited relevance of scientific explanation of the exterior phenomena and features of the process of evolution. The scientific theory of evolution is concerned only with the outward and visible machinery of the process of evolution. Matter and Life are indeed transitional terms facilitating the appearance of Man in the universe. But simply because Matter and Life are prior in point of time to Mind or Man, they cannot be regarded as superior in point of quality and reality to the latter.[8]

Sri Aurobindo acknowledges the principle of hereditary variation: 'Heredity upon which science builds its concepts of life evolution, is certainly a power, a machinery for keeping type or species in an unchanged being'. But he questions the adequacy of this principle in accounting for persistent and progressive variations. For, he thinks that the tendency of the principle of heredity is conservative or preservative rather than evolutionary.[9] Some biologists are of the view that each Grade of Being true to its type, exists in itself and by itself and has nothing to do with what precedes it or what succeeds it. This view is not acceptable to Sri Aurobindo. All that these biologists can claim on the basis of their findings is that the variability and actual variation of a type are confined within its own specification of Nature. And they have nothing to disprove the possibility of self-transcendence by a type of species. Sri Aurobindo is not in favour of regarding the existence of different grades or types of Being as the different resultant of the cumulative influence of hereditary variation. They are all due to the force of

Consciousness in the material universe. The practical or physical method of operation varies from grade to grade. But underlying the difference in the modes of operation there is a basic unity of Nature.

Nature starts her work in Matter with infinitesimals charged with immense energy, associating them by design and number. Energized infinitesimals are grouped together by the operation of the laws of Nature; group infinitesimals are the sensible objects of our knowledge. In the process of life formation, what consciousness-Force produces first are infinitesimal forms of life and infinitesimal animalcules.

> It creates an original plasm and multiplies it, creates the living cell as the unit, creates other kinds of minute biological apparatus like the seed or the gene, uses always the same method of grouping and association so as to build by its various operations and various living organisms.[10]

The creation of various types as such does not prove the case of evolution, for what is needed is an evolving and conscious purpose working in Nature. Sri Aurobindo holds that types vary within their respective boundaries; typal variation is not specific evolution, and a satisfactory explanation of specific evolution cannot be framed and offered exclusively in terms of heredity. Besides heredity, the forces which pave evolution are physical forces, such as food, light-rays and also perhaps some invisible forces of which our knowledge is still at a very rudimentary stage. The fact that some species under the influence of sub-conscious energy answer to the needs of environment, while others under the same influence remain unresponsive and fail to survive, is 'the sign of varying life energy and psychology, of a consciousness and the forces other than the physical at work making for variation in Nature.'[11] Typal variation, which is a symptom of a bigger internal movement, and trans-specific evolution are all being designed, engineered and successfully brought about by the force of consciousness. In other words, there is nothing blind and mechanical in the process of evolution.

Of course the evolution of Man may be explained without

teleology, but as Sri Aurobindo points out, in that case the supreme significance of man is denied. Teleology is generally abandoned in favour of two other possible modes of explanation. First, one can say with the materialist that all is the work of an inconscient energy which acts automatically by mechanical processes and have no element of purpose in it. Secondly, we may suppose that the Infinite and the Universal have everything in them already and that they have nothing more to realize, to accomplish or to work out. As we have already seen, Sri Aurobindo rejects both the scientific and the metaphysical denials of the teleological account of evolution. The scientific account of evolution is discounted by Sri Aurobindo, mainly on two scores. First, it fails to explain how evolution is different from involution; and, secondly, how specific (or in species) variation leads to the emergence of a superior type of being, and of an ideal form of human unity corresponding to that superior type of being. The positivist-minded scientist always seeks to explain *ought* in terms of *is*, ideals or possibilities in terms of *actualities*. His reductionist bias makes him suspicious of everything which is not immediately given or testable by what is given. He distrusts transcendental entities and even higher level scientific theories which have no verifiable empirical correlates. This sort of pseudo-scientific attitude towards science lands one in an uncritical pragmatism, makes one a worshipper of only whatever is given and sceptic about whatever is not given.

Sri Aurobindo reminds us that man is not a one-dimensional being. In our actualities we are true to our past and present, and through our possibilities we project ourselves to the infinite future. Man is indeed a project.

> Our actualities are the form and value or power of expression to which our nature and life have attained; their norm or law is the fixed arrangement and process proper to that stage of evolution. Our potentialities point us to a new form, value, power of expression with the new and appropriate arrangement and process which is their proper law and norm.[12]

In spite of this significant distinction between his actuality and his possibility or project character, man under the influence of his conservative reason is disposed to confuse the law of the future with that of the past and present, taking the latter as a sort of non-creative repetition of the former. He is more faithful to his actualities than to his possibilities, and that is why he, even at the actualities of his experience and environment, and even at his best can think only of that future which, in substance, is a mere reproduction of the past and the present. Man does not follow the spirit of the integral totality; he is not conscious of his heritage of the moving integral totality.

Creative consciousness is both integral and historical. The past is relived or is re-enacted in creative human consciousness. True historical consciousness does not mean the mere revival or repetition of what is past. If mankind partly forgets its past and gains new experience, that does not mean either the poverty or bankruptcy of modern culture. Similarly, if modern man lacks the integral outlook and the vision of the ancient seer, that does not mean that the modern outlook is necessarily inferior to the ancient one. In the place of epic vastness and classical grandeur there has been in later developments an increasing subtlety, complexity, manifold development of knowledge and possibility in man's achievements in his politics, society, life, science, metaphysic, knowledge of all kinds, art, literature. It is a new way of getting back to the past and exploring the future. Even in man's spiritual endeavour, Sri Aurobindo observes, this increasing subtlety, plasticity, sounding of depths and extension of seeking[13] are becoming more and more evident.

The process of evolution is then marked by a movement from simplicity to complexity and subtlety, from flexibility to rigidity and again back to plasticity, from inconscience through ignorance to knowledge. This movement encounters opposition and obstruction in its way and, as we have said before, is never linear. Even regressive or backward movement is not unknown in the course of evolution. Backward movements are only 'temporary descent' or at worst downward curves of the spiral of progress. Ultimately, Nature must move forward. Self-conscious man is the main agency and instrument of that movement. The evolution of man is destined to be progressive under the necessity of the Will of the Spirit. Through ups and downs Nature has a definite goal to realize in and through Man, transforming Man into Superman, transforming Mind into Supermind. "For (man's) cycles are circles of a growing, but still imperfect harmony and synthesis, and (Nature) brings him back violently to her original principles, sometimes even to something like her earlier conditions so that he may start afresh on a larger curve of progress and self fulfilment."[14]

Progress in fact is of two types—radical progress and adaptive progress. In the case of adaptive progress, there is no sharp departure from the past to the present; tradition, history and modernity are well balanced and preserved in it. In the name of radical progress, mankind in different times and places arrogated to itself the impossible task of making a clean sweep of the past, of starting from scratch, with an empty canvas before its eyes and a blueprint at its disposal. One may think of a man with no history behind him and society around him to determine and influence; but such a man is a theoretical abstraction and a practical impossibility. Because of an inner necessity, progress proves always adaptive, interpenetrative and rhythmic. These are all unmistakable expressions of the presence and operation of a purpose operative in history. The uniform rhythm of human cycles is a truth of the Spirit and being worked out by Nature. Disposing of the scientific and metaphysical objections against the teleological philosophy of evolution and history, Sri Aurobindo emphasizes the inescapable necessity of studying the human cycle from within, from the side of inner psychological forces and not from without, from the side of the outer material forces.

Sri Aurobindo does not believe in the autonomy of sociology, in the independent existence, action and reaction of social forces. According to him, they are all related to, influenced and supervened by the inner psychological forces. Spiritualist and teleologist as he is, he thinks that there is an inner harmony and synthesis between the two segments of social reality—psychological and material. While the material forces are not conscious of their origin, operation and direction, the psychological forces are, however, vaguely so.

The Marxist Objection : A Review

Sri Aurobindo's teleological philosophy of evolution and history is basically incompatible with the Marxist's account of the same.

Marx says that 'all historiography must begin from... natural bases and their modification in the course of history by men's activity'. By the 'natural bases' he means the geological, Pro-hydrographical, climatic and the life conditions of living human individuals. Sri Aurobindo tells us that the modification of natural bases referred to by Marx is brought about both in the human and in the sub-human grades, particularly in the latter by the conscious design of Nature. Even when man is found to be an active agency in bringing about modification and tansformation in the natural bases, he is acting knowingly in consonance with the purpose of Nature. This view is obviously unacceptable to Marx. After the natural bases, what a scientific historiographer must take into account are, says he, the sum of productive forces, capital and social relationships; and all these taken together should form the real bases of human history. Extending this line of argument, Marx affirms that the true conception of history rests on the exposition of real process of production, starting out from the simple material production of life, and on the comprehension of the form of intercourse connected with and created by this mode of production. The natural bases of history are transformed into real bases by man's own activity; and of the different forms of activity, Marx attaches greatest importance to the productive activity. He finds nothing divine, no design in this gradual and human transformation of Nature into social reality.

But, I think, it would be wrong to suggest that the materialist account of evolution as represented by the Marxist is concerned only with the external process, machinery and attending circumstances of evolution. To regard the human organism purely as a machine would perhaps be rash. And many Marxists would deny the criticism. Man is perhaps neither entirely a machine nor completely an individual. It is very difficult to substantiate the Vitalist's claim that human life is shaped and therefore explainable without reminder by its internal properties. In fact, we cannot isolate any system completely from the rest of the universe and definitely ascertain if it is completely mechanical or not. There is always an element of residue or uncertainty in the explanation offered exclusively either in terms of internal properties or in terms of external environment. It seems to be a part of the very nature of every form of life to interact with its environment. It not only adjusts and adapts itself to the environment, but also tries to adapt the environment to itself in some respects. It is very difficult to draw a sharp line of demarcation between an organism and its environment. It is undeniable that organic evolution is marked by a progress towards more and more complete individuality, both in the development of the individual and in the evolution of the race.

The Cartesian tradition is not yet dead. Descartes thought that the human body is a complex machine. Even those who, like Haldane, accept this mechanistic outlook deny that the behaviour and responses of the organism as a whole confirm the mechanistic hypothesis. They would rather say that '...the organism is something more than a machine, in the sense that it has a unity of a type which the machine lacks'.[15] Unlike the machine, there is something in the organism which enables it to repair itself and bring itself back to a normal state if it is not too seriously injured. But the ultimate subjection of all forms of life to death unmistakably indicates the superiority of the laws of material motion over those of biological motion.

Further, perhaps it would not be quite correct to mention the shortcomings of the Darwinian principle of Natural Selection in order to prove the untenability of the general materialist theory of the evolution of man. Perhaps we should remember that to justify his theory of evolution, Darwin spoke not only of Natural Selection but also of the inheritance of the effects of the use and disuse of organs by the organism. It is true that the latter view has not been as strongly defended by Darwin as Lamarck and Herbert Spencer. Consistently with his dialectical approach, Engels was more influenced by Lamarckism than by Darwinism. Darwin took variation, or rather heredity variation, for granted, and he did not seriously search for its origin. Over-impressed by the principle of heredity, he did not take serious note of the principle of mutation, which partially opposes and negates the work of heredity. Mutations generally disturb the normal relations between an organism and its environment and are, therefore, often regarded as harmful from the evolutionary point of view. Because of this, many biologists refuse to regard mutation as an evolutionary agency. But the problem is that qualitative novelty cannot be satisfactorily accounted in terms of purely natural selection, which is a negative process. Natural selection may at best account for the appearance of what is qualitatively new. But what makes qualitative distinction actually possible is something other than natural selection. It is not on this ground that in Marxist's quarters there has been a reaction against Darwin and a tendency to veer round to Lamarckism, a tendency to explain qualitative distinction exclusively in terms of the influence of the environment and its creative interference. It is to be remembered that Lamarckism itself has been interpreted in some cases very mechanically, and of course sometimes also in a creative way. Some Marxists, including Engels, betray the rather mechanical Lamarckian view that changes produced by the organism in response to a change of environment are directly inherited.

Haldane offers a different theory which tries to explain evolution in terms of a dialectical opposition of mutation and selection. And he claims that this is a Marxist interpretation of evolution, supported by modern researches in the field of biology.

Lysenko, the well-known Soviet scientist, also rejects the Darwinian theory of evolution, criticizing the principle of natural selection as 'flat' and pointing out its inconsistency with the conclusions of modern science. To him the classical theory of heredity (associated with the names of Mendel and Morgan) is also unacceptable. According to Mendel and Morgan, the bearers of inheritance are constituents of the cell nucleus, the chromosomes, or the genes contained in them. They explain changes of heredity, or mutation in terms of change that occurs in the constitution of the genes or chromosomes, which may in fact be achieved artificially. Lysenko does not believe that there are necessarily specific carriers of inheritance in the organism itself. His starting assumption is that organism and environment constitute a dialectical unity. A change in the environment offers man the opportunity of bringing about a change in the organism which can be passed on by inheritance to its descendants. To Lysenko heredity itself is not an abstract or a static principle. It is a concentrated form of the environmental conditions assimilated by the organism in a series of preceding generations. Usually the organism with a conservative heredity does not assimilate the conditions that are not native to it. But the organism with a de-stabilized or 'liberal' heredity accepts and assimilates the environmental conditions which environ it, with less discrimination, promoting greater intake and growth.

Lysenko criticizes the Darwinian theory.[16] He claims to have scientifically established the possibility of transforming one species into another. Unless the intra-specific composition of the organism is known, the desirable mode of inter-specific relation and the ideal, unity between organism and the environment cannot be correctly ascertained and established. The Marxist tries to prove that there is something in the organism which enables it to accept and assimilate some conditions of the environment and reject others. But he refuses to characterize or identify this native 'something' of the organism as spiritual. Further, the Marxist tries to give the impression that his dialectical approach should not be confused with the mechanical approach. While on ultimate analysis, he defends mechanism and determinism, and takes pains to point out that they do not clash with the spontaneity, initiative and creative response of the organism. In other words, the Marxist claims that the scientific determinism and creative enterprise of the organism are not at all incompatible.

Whatever might be the liberal interpretation of Marxism, one can hardly deny that the Marxist is very much concerned with the practical, particularly political, implications of his theory. Certainly one of the main reasons of his opposition to Darwin's theory of evolution is that in its sociological counterpart he finds an eloquent plea for unbridled economic competition. Influenced by Malthus, Darwin portrayed the struggle for life as an external struggle, a struggle between individuals rather than as a struggle between individuals and their environment, and as a struggle for external adaption rather than an internal struggle. Secondly, the Marxist rejects the Darwinian assumption that evolution is an uninterrupted and continuous process, believing as he does in a ceaseless class struggle climaxed by revolution.

Sri Aurobindo's theory also has a pronounced political implication. While Sri Aurobindo would claim that it is more or less a true statement of the unfolding of the Divine design in human history, the Marxist reiterates his position that there is nothing like Divine design and that human history is being created and modified by man himself. Yet, man himself being subject directly to economic laws and indirectly to those of Nature, his freedom to create and modify history is always a matter of degree and never absolute. Regarding man's position vis-a-vis the Divine design, Sri Aurobindo points out that there is nothing which might suggest that man is not free to shape his own destiny. His place in Nature does not stand in his way of a search for greater ideals and larger freedom. Man wants to be wholly conscious of his being. Self-consciousness is what is meant by spiritual knowledge and its essence is an intrinsic self-existent consciousness. But here again man's subjection to the laws of Matter and Life enables him only to be partially (and not integrally) conscious of his total being. Both Sri Aurobindo and Marx agree, in different ways though, that man is the author of his socio-historical destiny, but not solely.

But the point of agreement must not make us blind to their fundamental differences. The Marxist thinks that the metaphysical presuppositions of Sri Aurobindo's sociology are not only superfluous but also scientifically untenable. An Aurobindite argues that the Marxist account of the human cycle is superficial. Despite his denial of the existence and operation of an intelligent principle underlying the process of history, the Marxist admits that the histories of different societies, separated from each other by time and space, exhibit certain definite rhythms, indicating thereby that there are some general laws of history. Sri Aurobindo also finds definite rhythms in history; but believing as he does in the existence and operation of an intelligent principle in history, it is with greater theoretical plausibility that he can speak of universal patterns or rhythms of history. According to him, the materialist conception of history is superficial and reveals empirical trends sustained by a deeper teleological movement of Nature. He finds no incompatibility between the deeper natural laws of history and superficial empirical trends. In defence of his own view, he argues that the Marxist's commitment to the empirical ('dialectical') theory of knowledge makes it impossible for him to go deep into the workings of human history and discover its persistent inner laws.

REFERENCES

1. *The Life Divine* (LD), pp. 659-60, 1 Vol. edn., New York, 1949. See also,

Pierre Teilhard de Chardin, *The Phenomenon of Matt,* Fontana Books, London, 1971.

2. F. A. Hayek, *The Counter-Revolution of Science,* 1964, pp. 17-24.
3. *The Human Cycle* (HC), Sri Aurobindo Ashram, Pondicherry, 1949, pp. 292.
4. LD, p. 737.
5. LD, p. 743.
6. *Anti-Duhring,* p. 115, Moscow, 1954. See also, Jacques Barzun, *Darwin, Marx and Wagner: Critique of a Heritage,* 2nd ed., New York, 1978.
7. LD, p. 662. See, also Jack Lindsay, *Marxism and Contemporary Science,* London, 1949.
8. LD, pp. 744-45.
9. LD, p. 738. See also, Julian Huxley, *Soviet Genetics,* London, 1949.
10. LD, p. 739.
11. LD, p. 740.
12. *The Ideal of Human Unity,* Sri Aurobindo Asharm, Pondicherry, 1950, p. 179.
13. LD, p. 749.
14. HC, p. 98.
15. J.B.S. Haldane, *The Marxist Philosophy and the Sciences,* p. 104, London. 1939. A critic of Lysenko, he was deeply influenced by the theory of genetic mutation due to De Vries.
16. *Dialectical Materialism: A Historical and Systematic Survey af Philosophy in the Soviet Union,* Gustav A. Wetter, translated by Peter Heath, pp. 455-69, London, 1958. See also, C.H. Waddington, *Introduction to Modern Genetics,* London, 1950.

4

PIERRE TEILHARD DE CHARDIN :

AN EVOLUTIONARY VIEW OF MAN

The best tribute to a great thinker is to try to understand critically his ideas, remove the attending misunderstandings, and develop in different ways, intended as well as unintended, the basic insights underlying those seminal ideas.

Pierre Teilhard de Chardin (1881-1955) and his main work, *The Phenomenon of Man,* which unfortunately could not be published during his life time are equally extraordinary. The author was a passionate truth-seeker and the works left behind by him are full of daring ideas controlled, to a great extent, by scientific findings and arguments. If at times Pierre Teilhard's religious conviction seems to have prevailed over his carefully cultivated scientific temper, it is mainly because of the impossibility of the total absence of the author from his work. Utter objectivity or impersonality is a widely entertained myth.

That man is a transitional middle term in the course of evolution with a pre-human geological and biological past behind him and a super-human future ahead of him has been acknowledged by many thinkers, theologians and scientists, of quite different persuasions, anti-naturalist and pronaturalist. Their idioms and formulations differ because, among other things, of their different social, scientific and religious contexts. Even the static structuralists, who reject the very notion of evolution and fall back on a dualistic position, can hardly deny the gradual and qualitative enrichment of the human species as evident from massive and diverse observational data of diferent life-sciences. And their caveat that these evidential data being superficial in nature do not reveal the true structure of man and, therefore, need not be taken seriously, betrays only their persistent transcendental bias, disregarding ultimately the findings of science. What particularly impresses one is that, in spite of his Christian theological conviction and transcendental vision of Omega-workings, Pierre Teilhard is seriously engaged in a critical dialogue with the physical and life scientists to settle his account with them.

The story of human evolution, as we know, is very old and varied. The Indian thinkers of the Vedas and the *Upanishads* repeatedly returned to and discussed the theme, of course, in their own cryptic-symbolic ways. "His ecstasy, in one who holds it, sets into motion the births, the human self-expression and the divine, and moves between them."[1] "Masters of the Truth-Light who make the Truth grow by the Truth."[2] "The Individual is in the Universal; the Universal in the Individual; the Individual and the Univeral are identical at bottom." The insights of the early Indian thinkers, like those of their European counterparts, have been developed in two different directions, pro-naturalist and anti-naturalist, the *Sankhya* and the *Vedantin,* for example. In the recent past an Indian contemporary of Pierre Teilhard, Sri Aurobindo (1872-1950) has developed an evolutionary system of integral *monoism, Purnadvaitavad,* a pro-Yedantin system ingeniously integrating the pro-naturalist elements of *Sankhya* and *Yoga* in it.[3] Sri Aurobindo worked out his system by 1920, and, it seems, Pierre Teilhard arrived at his conclusions quite independently. One is bound to be struck by the similarities of their approach and the conclusions. Already some comparative studies of the ideas of these two thinkers have been made.

Among the European thinkers, Aristotle was perhaps the first who made a very systematic study of *development*—the properties of natural bodies and how those are related to the properties of biological bodies. In physics he studies movements, forms and transformations of natural bodies (both terrestrial and celestial) *in general,* and in *biology* those of living bodies *in particular.* In the Aristotelian scheme of thought the movement is from *general* to *particular.* He noted that every body is an "embodiment of" and every piece of matter, living or non-living, is informed. The main feature of his view of the Universe is his notion of teleology: in nature there is an unconscious striving towards ends. God and nature do nothing in vain. There is nothing in the universe without a purpose. *(De An.* 1075al5, 1076al4, 1075a19).

Lucretius was no believer in teleology. His main contribution to philosophy has been his pro-naturalist account of human evolution. The human species survives, according to him, not because of any calculated pursuit of self-interest by its members but because of their native sense of justice and a natural instinct working in each person impelling him to cooperate with others. All things, human and natural, develop by degrees with the passage of time and this process goes on in an orderly manner "until these things have reached their highest point of development." *(On The Nature of Things,* Bk V, 1448).

It is very interesting to note that not only the ancient pro-naturalist thinkers like Lucretius, Epicures, or the authors of *Sankhya-yoga* systems of India but also their modern successors, with vast informations of natural life and ecological sciences at their disposal, are disinclined to accept the human mind, as we know it today, as the omega term of the evolutionary process. In spite of his pronounced positivism, Comte came out in defence of a "Religion of Humanity" and was for developing an attitude of reverence towards *Humanity,* a collective *mystique,* if not a God-surrogate. Footfirm in such experimental sciences as physics and physiology, Samuel Alexander did recognize the irreversible possibility (Eddington's "time-arrow") of the emergence of complex structures of mental properties out of spatio-temporal matrix and which would deserve the general designation deity (in no metaphoric sense). It seems to me that when the full implications of the following two stages of Marx's notion of the future Communist Society are objectively worked out without ideological prejudice, it would be clear that he also thought of creating both subjective and objective conditions making "the development of human potentiality for its own sake" possible. One : "the socialized mankind [would] regulate their *interchange* with Nature rationally, bring it under their common Control. . . and accomplish their task with the least *expenditure of energy* and under such conditions as are proper and *worthy* for human beings"; and, two : "Beyond [this realm of necessity] begins that development of *human potentiality for its own sake,* the true realm of freedom... [with the] realm of necessity as its basis." [italics mine](Capital III). Given the irreversibility of time *and* dialectical interchange between nature's powers and the socialized mankind's, I do not quite see how the consistent Marxist can escape the conclusion of Spinoza or for that of Einstein viz. a unified deterministic system which is essentially free from within and grows by that power of freedom (observationally exercised only by human beings severally or jointly). If Spinoza could be a mystic and pantheist to Goethe and Coleridge a monistic materialist to Plekhanov, a dialectical emergentist and pro-freedom interpretation of Marx can hardly sound surprising. Rather, to my mind, it accords well with the basic humanistic orientation of Marx's thought.

Man has always been the most fascinating subject of his own study. He can study himself as an object, one among other objects of the world. Objections have been raised, among others, by the Vedantin and the Kantian that "self as an object" cannot be an object of valid knowledge. For the knower is essentially and necessarily the subject of knowledge. The image or idol of self, the object of possible knowledge, is not self in the true sense.

From the ancient alchemist to the modern biochemist or ecologist, whoever has tried to know man has been led to know his place in nature, what elements of the environment have gone into the making of his nature, and what influences of the environment have entered into his psychological make-up, dispositions, shaping his destiny. Man is really environed by Nature, both physical and social. Nature, in turn, is also being shaped, distorted, damaged, exploited by man. But that these processes of interaction are not simply causal, linear or curvilinear, and becoming increasingly complex, cybernetic, involving feedback, and concentric was not clearly and comprehensively realised earlier.

Phenomenon is something that shines, unfolds itself and illumines others at the same time. Pierre Teilhard has shown us in a very masterly manner that man is the phenomenon of phenomena, in which enfolded nature unfolds itself, and, to quote Julian Huxley, "evolution becomes conscious of itself." The highly readable and the lucid style of his expression might conceal his remarkable scholarship from the uninitiated in the vast area of the subject of his inquiry. The competence of the thinker and the quality of his works could be gathered from the fact that even professional scientists and philosophers, including those who do not share his religious views or transcendental sweep of thought, agree that he has faced the phenomena, big and small, from sub-atomic particles and cells to stellar galaxies, biospheric and noo-spheric phenomena, with a deep intellectual love and tried to see them as integral parts of an evolving system. Julian Huxley and Joseph Needham, both wellknown for their scientific contributions and philosophical orientation, have commended his system in rare complimentary terms.

Within the limited scope of this paper I propose to indicate few basic features of his system and raise some questions which seemed to me central to a proper understanding of Pierre Teilhard's main conclusions. My choice of the features of his system has obviously been influenced by the questions I have had in my mind.

Evolution is the key concept of his thought. The whole of knowable reality and every aspect of it—matter, life, mind, history and values, not excepting the knower himself are viewed not as a static unity but as a developing process marked by increasing complexity. His preferred terms, like, *cosmogenesis, hominization, noogenesis,* and *ultra hominization,* have their definite connotations in his system and are critical stages of the forward looking evolutionary process. The powers and properties of the different levels—material, vital, mental and super-mental —are not entirely different; on the contrary, they interpenetrate, resulting in increasing complexity and concentration. This evolutionary process cannot be adequately described or evaluated in terms of its origin. Thus he avoids what is called genetic fallacy: by tracing the genesis of a phenomenon that canot be explained. Certainly the human phenomenon, which according to Nietzsche is still incomplete, is not amenable to genetic or causal explanation. Purposive explanation is called for. The more developed mental phenomena, for example, cannot be satisfactorily explained by the less developed ones. Because that would involve fallacious reductionism, leaving the qualitatively richer, newer and emergent noospheric levels partly unaccounted for.

The whole process of evolution from subatomic units to atoms, from atoms to inorganic and organic molecules, thence to sub-cellular self-replicating assemblages of molecules, and then to cells, to multicellular individuals, to cophalized metazoa with brains and protomentality, to primitive man, and civil society, are characterized by two apparently contrary but functionally complementary tendencies— *differentiation* and *integration.* This twin tendencies may also be designated by one common name, *self-complexification* or "convergent integration". Nature does not encourage endless proliferation of entities and groups and evolution involves their unification. This is clear from all levels of evolutionary nature, from the inorganic to the social. And this is due to simultaneous operation, in fact cooperation (marked by subsidiary conflict), of two modes of energy, measurable-material and immeasurable-psychic. Every phenomenon seems to have two aspects, "physoid" (nature-like) and "mentiond" (mind-like), and exchange of powers and properties between them account for the emergence of new powers and properties in the concerned phenomenon and its increasingly complex metamorphosis. Whatever it is, physical particle or living cell or human society, the two aspects are there, one responsible for its relative stability and another for its relative mobility. Their ways of being and working are creatively interactive, and not parallel upto a point and then convergent as Spinoza was inclined to believe. However, in the field of cultural anthropology, Pierre Teilhard defends parallelism and rejects diffusionism: but only upto a point, a critical point, and not beyond that. Cartesian or Sankhya-type of dualism is quite alien to his thought; even Kant's proto-teleology is not sufficient for his theoretical purpose of projecting the whole mankind as a phenomenon. He rejects dualism at all levels— wave-particle, DNA-RNA, body-mind, and so on. Each one of these pair of entities is informed of and informing the other. He speaks of two forms of energy—physical and psychical. "Between the *within* and the *without of* things, the interdependence of energy is incontestable. "[4] *Essentially* all energy is said to be psychic. But Pierre Teilhard's functional formulation of its two components is very sophisticated and purported to buttress his evolutionary theory; a *tangential energy* relates the concerned entity with all others of the same order, of the same complexity and the same centricity; and a *radial energy* draws it towards even greater complexity and centricity, breaking through its involutionary or enfolded cycles, and imparting in it an evolutionary or unfolding and a more encompassing thrust. At the University of California in 1952, Julian Huxley tells us that Pierre Teilhard "found" a parallelism between "the cyclotron generating immense intensities of physical energy in the inwardly accelerating spiral orbits of its fields of force, and the entire noosphere with its fields of thought curved round upon themselves to generate new levels of "psychical energy."[5] Huxley wonders "how his imagination would have kindled at the sight of the circular torus of Zeta, within whose bounding curves are generated the highest physical energies ever produced by man!"

It is indeed very interesting to note that without any professional familiarity with science and modern scientific machines, Sri Aurobindo relying mainly on the intuitively grasped first principles, made transparent by in-depth reflection and ratiocination, brought out their enormously complex implications, and could also clearly see through the emergent evolutionary role of what is called existence as energy. He speaks of —

> self-aware force of existence of which mentality is a middle term; below mentality it sinks into vital and material movements... above, it rises into the supramental... But in all it is one and the same thing organising itself differently. This is... the Indian conception of *chit*, which, as energy, creates the worlds. Essentially, we arrive at that unity which materialistic Science perceives from the other end when it asserts that Mind cannot be another force than Matter, but must be merely development and outcome of material energy. Indian thought at its deepest affirms on the other hand that Mind and Matter are rather different grades of the same energy, different organizations of one conscious Force of Existence.[6]

The main work of Sri Aurobindo was written nearly three decades before that of Pierre Teilhard. I know of no evidence that the latter was aware of the former's thought. Only towards the end of his life Teilhard came to know of Sri Aurobindo's views on the human evolution. Yet the similarities of their central theses are not only striking but also engagingly educative. Pierre Teilhard's twin conceptions of *tangential energy* and *radial energy* are essential to the creative dialectic of self-arming involution and self-exceeding evolution. "Behind the action of the material Energy" Sri Aurobindo speaks of the presence and workings of "a secret involved consciousness, cosmic, infinite." Material energy is said also to be "frontal energy". It is "an indispensable condition for the structure of the material world-substance, in which this consciousness intends to involve itself so that it may grow by evolution out of its apparent opposite; for without some such device a complete involution would be impossible." In Sri Aurobindo's language (and here he differs from Sankara's Vedanta), Consciousness is Consciousness-Force and it has two modes of manifestation, (a) frontal (material) energy, and (b) substantive (spiritual) existence. The former is knowable only by sense-organs and scientific methods, the latter by the subtler or non-sensuous powers of the mind. Since there is no division between these two aspects of Existence, the forms of material energy expresses, though inarticulately, the secret troths and ways of working of the inner spiritual existence.

Sri Aurobindo's theses on biology, psychology and sociology are all integrally related by his main evolutionary metaphysical theory. Compared to Pierre Teilhard's language, Sri Aurobindo sounds more metaphysical. But in fairness to both it has to be admitted that their thoughts have not only taken serious note of the contemporary science but, what is more constructive and courageous, they have opened their metaphysical conclusions to the tests of empirical sciences. Critical dialogue between science and metaphysics saves the former from sterile positivism and the latter from inconsequential empty speculation.

I for one am sympathetic to metaphysical, speculative and transcendental approach to sciences, facts of life, in order to make them systematically intelligible and show their close relation to the needs and queries of even ordinary life. The fall of logical positivism and the decline of analytical philosophy in the recent years are primarily to be attributed to the overzealous proponents' refusal to face the larger issues of the human life, viz., the place of mind in nature, the presence of values in the world of matter and life, the yearning for larger human aggregates and unity, the proclaimed omnipotence of God and the evil, the survival of the religious instinct of the common man in spite of the marvels of science and the general critical attitude of the professional intellectuals, the evolution of mankind—a new super-human species, and the thermodynamic forecast of the cold-death of both the noosphere and the biosphere. Metaphysical issues do have their undoubted charm and yet raise a host of difficult questions.

My first difficulty with Teilhard is that he takes Man (with a capital M) and a, i.e. one single, phenomenon. It is one thing to discuss scientifically biosphere or noosphere and its properties in a holistic manner for a limited heuristic or expository purpose, but it is a quite different thing to say and also mean that mankind in its totality is a phenomenon and that it can be described and analysed like any other phenomenon. It is true that the biologist and the epistomologist speak of some biological properties and epistemic propensities common to all human beings, irrespective of their geographical habitation and socio-cultural affiliation. In the context of yesterday's Nazi Biology and today's talk of lethal nuclear weapons, war and tension, it is not at all surprising that philosophically disposed biological scientists—Huxley, Haldane, Waddington and Sherrington, e.g., would bring out the normative implications of life-sciences, stressing the unity of mankind. Haldane observes, "God is now enlarging the sphere of human choice. . . giving us new duties" (to remodel the human society on the basis of available scientific knowledge).[7] Sherrington tells us, "Ours is a situation which transforms the human spirit's task. . . to one of loftier responsibility. . . (raising) the lowliest human being conjointly with the highest, Prometheus-like, to a rank of obligation and pathos which Moses in his law-giving nor Job in all his suffering could surpass."[8] Advocating his evolutionary humanism Huxley urges upon us to develop a "global out-look" and thinks that "Religion of some sort is probably necessary."[9] None of these thinkers perhaps would share Teilhard's profound spiritual and religious belief that mankind is *essentially* one. *Essentialism,* I suspect, often conceals untenable metaphysics and, what is more disturbing, makes it impossible to look critically into the grounds of untenability. The case of Teilhard, I gladly agree, is not quite of that sort. Even then my suspicion lingers on. Gathered from different cultural disciplines, we have now at our disposal mass of facts to show that human peculiarities are various, persistent and almost irremediable. The main defence of freedom rests on the peculiarities of individual human beings. If all of us are deemed to be sands in the sea, then in the name of "compression" our "expression" is bound to be berated. It is true that Teilhard has tried to defend the notion of freedom at many different places of his published works and particularly through such concepts as "centricity", "complexification" and "expression". Had he been opposed to freedom and a defender of barren uniformism, it may be argued, he would not have taken so much of pains to elucidate these concepts. I agree. But, at the same time, we must not fail to note his "relentless insistence" on the immersion of our body and soul in the irresistible stream of space-time,[10] on the ultimate unity of Man and Mankind and the ceaseless process of collectivization of Man in Mankind and in-folding of Mankind in Man,[11] and, also, on the Omega-governed destiny of Man.[12] For Teilhard freedom is a strong form of recognition of necessity. For him the locus of true freedom is Mankind and not the individual Man. Man, then, is not free in his individuality, but in the concentric unity of mankind. Can a time-swept individual, not himself being free, be rationally supposed to be free in the complexified community of like unfree individuals? Or, should we think Mankind knows no unfree individuals in it? I feel the difficulties, particularly when I think of individuals like me, you or him.

Denial of holism in anthropology (or science of Man) is bound to have its consequence in epistemology. If Man is same as Mankind (speaking in terms of structural unity or potentiality), why are we so differently informed by our environment, and why those of us who are uniformly informed exhibit in our thoughts and actions varying degrees of uncertainties and anxieties? If our cognition (or lack thereof) of our world is so different, what is the compelling reason, if any, for supposing that Mankind is infolded in individual human beings? We differ so much in our thoughts and actions, in our perception of "the" world in and around us, one wonders, whether we should still maintain that Man in Mankind is one uniform phenomenon. A similar view has been defended by George Sarton in another but related context of the history of science. The fundamental unity of life, says he, is exhibited by its three aspects, viz., (i) nature is one, (ii) science is one, and (iii) mankind is one.[13] I admit that in principle we are entitled to do so. But, I suppose, this admission is a near-empty theoretical pronouncement; this principle is metaphysical (i.e. uncriticizable by any *testing* empirical evidence); and, therefore, the said title is very weak. The *supporting* facts and evidences marshalled by Teilhard and Sarton are highly selective and interpretative. They seem to be more interested in *support* for and not *test* of their view.

Collateral to, or, one might say, as a part of, the above difficulty I have another problem with Teilhard. His claim that mankind and all its manifestations are "proper objects for scientific study", taken together with his assertion that scientific "discovery and synthesis are no longer merely speculation but creation", is at the root of my problem. Science has no doubt a value of its own but it is needed and solicited mainly because of its power—power of "collective vision obtained by a panhuman effort of investigation and construction." It is again interesting to recall Sarton on this point: "the history of science is the history of mankind's unity, of its sublime purpose, of its gradual redemption."[14] My uneasiness with these expressions are not mainly linguistic but substantive. True, knowing involves making, creating or framing (of hypotheses and concepts, etc.), but if we overemphasize, as Teilhard does, this aspect of scientific knowledge we treat may be unintentionally sciences as *instruments*. The instrumentalist, as we know, is inclined to change the image of science: for him it is primarily a means to realize some end which may or may not be cognitive in character. For Teilhard science is the basic hope-fulfilling enterprise—hope for the Omega-unity of Mankind. Practical use of science is of course welcome: it is even necessary for preserving the critical and growing character of science. But to harp on the use-value of science, particularly when the concerned value is emotive (even if in the good sense) and not cognitive, may lead to misuse of science (in the name of science).

My last and main difficulty in following Teilhard's otherwise fascinating account of Man's evolution pertains to its super-life phase. Mankind is destined to move beyond its present level and, it has been argued, *entropy* does not threaten in the least its existence on the earth. For other equally potent factors, viz., *expansion* (by which the layers of the universe unfold and granulate), *electrical and gravitational attraction* (by which sidereal dust conglomerates), and "interiorizing complexification" (which animates the whole mass of things) are said to be simultaneously in operation, resulting in increasing external complexity and internal concentricity of consciousness. On the implications of the last factor, Teilhard concedes, "science has not yet said its last words."[15] However, that is not a very important point by itself: a philosopher, a visualizer, need not follow science in servile manner, abandoning his own cultivated intuitive power or, what Einstein used to call, love of reality. The cause of difficulty lies elsewhere.

The concepts of *ultra-hominization, hyper-personalization, megasynthesis* and similar others are certainly very insightful and new in formulation; I say "in formulation" because from Nietzsche to Sri Aurobindo several other philosophers, visualizers offered kindred concepts—*overmind, supermind, gnostic being, supramentalization, superman, Divine Life,* etc. It has also been repeatedly claimed that the seers of the ancient times had left behind analogous "luminous ideas". I do not propose to contest this claim. I myself believe that there are different, mystic or even mysterious, ways of knowing. But once we start saying that what we know in paranormal or intuitive ways are ipso facto valid, i.e. self-certifying, and need not be subjected to the scrutiny of sense and science, we would be obliterating the respectable, may be changing, line of *demarcation* between serious scientific inquiries and pseudo-scientific divinations or lesser type (though marginally empirical) inquiries.

Compared to many other religious-minded metaphysicians Teilhard's familiarity with sciences is much deeper and he has made perceptive use of available scientific findings in formulating his daring vision of the future of Man. But I have a persistent feeling that the visionary metaphysician in him has proved more assertive than rigorously reasoning scientist in him. Partially supported by scientific theories, he rushes to some very bold and intriguing conclusions regarding the super-life of Man. By itself this approach, to my mind, is perfectly scientific. But my complaint is that Teilhard has not exposed his science-inspired conclusions to the logical scrutiny of other equally science-loving (may be less religious in disposition) philosophers and thinkers.

Let me try to make my point more specific. His claim that the noosphere will be intensely unified and attain a hyperpersonal organization (i.e. Omega point) is anti-entropic, running contrary to the second law of thermodynamics with its dissipation of energy and its tendency to uniformity. To meet this criticism the facts and the considerations which he marshalls show at best *only the possibility* that even in an entropic ocean there may survive or emerge few islands of life. After emergence, on his own showing, there comes emersion. But even then the super-life of the spirit, we are told, is bound to re-emerge. The morale of the resurrection, carried too far, makes little scientific sense. On the insufficient basis of anti-entropic possibility of the survival of life the superstructure of the super-life, even if constructed, cannot be maintained for long. Merely on the supposition that Man could not have evolved to remain for ever what it is, *knowledge* cannot possibly *create a* hyper personal Mankind capable of defying cold-death. From the scientifically available accounts of geology, palaeontology, biology and psychology (including parapsychology), one is not justified to infer either deductively or inductively the Divine future of Mankind. Not deductively because the emergent properties are not contained in the premise-principles. Nor inductively either, because the theoretically necessary presuppositions are not obtained, and, besides, the emergent properties of the Omega point are *not logically projectible.* Human hopes, prayers and even collective will are not enough to bring about that supreme stage of the evolution of Man. It is difficult to deny that many of us hope and will that the human existence, both individual and collective, be transformed into a qualitatively rich state, marked by every sort of perfection. But all that I doubt is whether there is *reasonable* ground of this hope and will.

When I speak of certain difficulties in Teilhard's Philosophy of Man, I do recognize the merits of his remarkable contribution. May be if I study his works once again, some of these difficulties will be removed from my mind. Who knows the evolutionary nature of knowledge of the universe will not raise other difficulties?

REFERENCES

1. Rg Veda, IX. 86, 42.
2. Ibid., I. 23.5.
3. D.P. Chattopadhyaya, *History, Society and Polity: Integral Sociology of Sri Aurobindo,* Macmillan, New Delhi, 1976.
4. Pierre Teilhard de Chardin, *The Phenomenon of Man,* Fontana, London, 1971, p. 70.

5. Julian Huxley, *Essays of a Humanist,* 1954, p. 207.
6. Sri Aurobindo, *The Life Divine,* Birth Centenary Library, Vol. 18, p. 88.
7. J.B.S. Haldane, *Everything Has a History,* 1951, p. 287.
8 . Charles Sherrington, *Man on His nature,* 1975, p. 294.
9. Julian Huxley, *op. cit.*, p. 87.
10. Pierre Teilhard de Chardin, *op. cit.*, pp. 154, 242-44.
11. ——, *The Future of Man,* Fontana, London, 1973, pp. 133,190--9I.
12. ——, *Man's Place in Nature,* Fontana, London, 1971, pp. 120-21.
13. George Sarton, *Introduction to the History of Science: From Homer to Omar Khayyam*, Baltimore, 1927, p. 31.
14. ——, ibid., p. 32.
15. Pierre Teilhard de Chardin, see (12) above, p. 35.

5

Sri Aurobindo and Pierre Teilhard de Chardin on Human Evolution

I propose to look into some points of agreement and difference between Sri Aurobindo and Teilhard on human evolution. Further, the question of the relation between entropy and life deserves a closer scrutiny. Also, the larger related question of the relation between science and religion, seems to me, merits critical reviews.

I
POINTS OF AGREEMENT BETWEEN SRI AUROBINDO AND TEILHARD

Evolutionism against Structuralism

Both Sri Aurobindo and Teilhard are evolutionist and oppose the static-structural account of reality. The former, primarily following the Vedic and the Upanisadic insights and influenced by his own historical studies, develops a theory of evolutionism which is fair to the facts of contemporary physical and life sciences and yet remains basically monistic in its orientation. Teilhard, on his part, was a paleontologist, a professional scientist, familiar with different natural sciences, and carried out extensive field research. He does not claim himself to be a philosopher. Both are deeply religious persons. Though himself a Jesuit priest, it is interesting to recall in the twentieth century, Teilhard's works were not approved of by the Vatican authority and therefore remained unpublished until his death. Sri Aurobindo's and Teilhard's *anti-structuralism* is grounded on their common conception that the world of sense-experience is real and that man's abilities to know the world are not identical throughout. In the course of time the properties of Man undergo significant change and, therefore, what the world is like cannot be ascertained in terms of one unalterable set of concepts. The meaning of the very concept of property is theory-bound.[1] What intelligence of ape means in zoology is quite different from what it means in human psychology or computer science. "Life" means one thing to the *Carvakas* and another in the systems of Sri Aurobindo. This point has been rightly emphasized by Nagel[2] to clarify the issues involved in "the reduction of theories", of biology into physics.

According to Teilhard, a philosopher wants to give us a coherent picture of the world "intuitively endowed with certain fixed properties which are not a thing in themselves but a general condition of knowledge." Since these properties change, Teilhard argues, the relation between the parts of the coherent picture of the world given by the philosopher also undergoes change. The change in properties, however, is to be understood with reference to the theories concerned and not in isolation. "Indeed the past history of human intelligence is full of 'mutations' of this kind. . . "[3]

A similar line of argument is found in Sri Aurobindo's writings as well.[4] Knowledge may be conceived as a phenomenon in and of reality itself. It is reflective of self-conscious part of reality itself. In and through Man's knowledge, reality comes to know of what it is. To our knowledge, reality does not owe its being. It is the other way around: reality makes mind, mental and supra-mental knowledge possible. By implication, noetic dualism is denied both by Sri Aurobindo and Teilhard. But, as we know, through common sense and science, knowledge is conceived as an achievement of man as a knower about something which is known, the object of knowledge, and which is separate from the knower. Sri Aurobindo argues at length to show different degrees and types of separation and identity between the knower and the known. Separation may be complete or partial. Sense-knowledge is more separative and intellectual one less so. Similarly, identity may be partial or complete. In intuition of the mental plane, the knower becomes partially identified with the known. In supramental intuition the identity between the knower and the known is completely realized. The ideal form of knowledge, knowledge by identity, according to Sri Aurobindo, is not achievable in terms of empirically available concepts. While, for the purpose of organizing the objects of sense-experience, he recognizes the necessity of empirical concepts, he takes pains to show their inadequacy and to highlight the significance and necessity of intuitive informations of the world, identificative knowledge, the knowledge in which reality is immediately reflected. In brief, Sri Aurobindo and Teilhard plead for the acceptance of alternative conceptual frameworks to deal with different spheres of objects—the physical, the vital, the mental, and the supra-mental. These alternatives are both functionally and evolutively, finely graded.

Principles of Convergence and Divergence

Sri Aurobindo and Teilhard think that the process of evolution is marked by increasing complexity and concentration of the elements evolving. The higher the sphere of evolution, the more intensive is the "complexification" and "concentration." Sometimes it is said to be the self-involution character of the evolving world. Life is more complex than matter and, analysis reveals that mind is more complex than life. And it is asserted that the super-mind would be even more complex. Teilhard frequently uses expressions such as "complexification of matter," "centro-complexity," increasing "convergence" and increasing "interiorization." Divergence of processes is followed by their convergence. First comes unfolding, then enfoldment. The latter phase retains the properties of the former phase in a concentrated form and adds something more to it. Teilhard's thesis is that "growth of consciousness" is nothing but "manifestation of organization." Higher consciousness involves higher organization or increased interiorization of psychism of "cerebralization".[5]

It is interesting to note that Sri Aurobindo characterizes the evolutionary process by *ascent* and *integration,* and claims to have taken the cue of his thought from the Vedas. "I have arisen from the earth to mid-world . . . from the mid-world to the heaven, from the firmament of heaven I have gone to the Sunworld, the Light."[6] The earth is Matter, the mid-world Life, the heaven Pure Mind, and the Sun-world Supermind. That is how Sri Aurobindo interprets the metaphoric terms forming an ascending series, marked by increasing organization, integration, enrichment of properties and power. "An evolution of forms of Matter more and more subtly and intricately organized so as to admit the action of a growing, a more and more complex and subtle and capable organization of consciousness is the indispensable physical foundation.... An upward evolutionary progress of the consciousness itself from grade to grade, an ascent is the evident spiral line or emerging curve that, on this foundation, the evolution must describe."[7] Elsewhere Sri Aurobindo speaks of "process of heightening and widening and integralization," "unification" and "centration", which characterize the process of evolution, throwing up higher and higher forms of consciousness.[8]

It may be recalled at this stage that evolutionary concepts came particularly in vogue in the middle of the nineteenth century. Herbert Spencer in his monumental works on biology, psychology, sociology, and ethics, has tried to develop a pro-naturalist theory of evolution. His concepts and expressions anticipate those of Lloyd Morgan, Sri Aurobindo, Samuel Alexander and Teilhard. According to him, "evolution is an integration of matter and concomitant dissipation of motion, during which the matter passes from a relatively indefinite, incoherent homogeneity to a relatively definite, coherent heterogeneity; and during which the retained motion undergoes a parallel transformation."[9] Even the forms of sociation found in the human life are traced by Spencer into biological level in the life of animals and insects. Not only mentally, but also biologically the human species is equipped to develop higher forms of association and organization. Spencer finds no incompatibility between scientific truths and religion, for, with him, both are expressions of one Omnipotent Power. A similar line of argument is there in Samuel Alexander. Alexander's *Space, Time, and Deity* is an extended legacy of Kant and Hegel and, at the same time, anticipation of Teilhard. Influenced by Lloyd Morgan's concept of emergent biology and psychological principles of Munsterberg, Alexander develops a pronaturalist theory of evolution and claims that his deity is not religious in any ordinary sense. However, he admits that the religious and moral feelings of the living beings might be construed as the *nisus,* an upsurge facilitating the process of evolution, and that the universe is "pregnant" with the qualities of beings not yet empirically available. Alexander uses the Kantian categories of substance, cause, number, and relations as basic, but following Einstein, who in turn himself was influenced by Kant, speaks of space-time as a four-dimensional continuum. Kant spoke only of the *functional unity* of the spatial and the temporal. It was left to Hegel to inject the notion of *continuum* in the Kantian *structure*. Alexander, like Spencer, claims that his evolutionary theory is also pro-naturalist and anti-teleological. He characterizes space-time as the constitutive stuff of the developing world, "pregnant" with the qualities yet to evolve. From this basic premise he derives the theorem that time is an attribute of reality and is mental in character. For him knowledge is a situational "compresence" of object ("contemplated") in a subject ("enjoying" it). From their own different points of view Sri Aurobindo and Teilhard think that knowledge is a situation of a phenomenon in reality, and not to be understood as an exclusive human achievement. Knowledge is made possible not merely by the mentality of the human beings, but also other non-mental factors, for example the physical and the biological ones, which are working continuously in and around the human mind.

All these evolutionists do recognize that not only psychologically but also biologically the human species is equipped to be a social being, capable of forming and developing society.

Inevitability of Socialism, Its Inadequacy, and Transcendence

Extending their main theories of evolutions from life to the level of mind, the noosphere, Sri Aurobindo and Teilhard speak of the inevitability of socialism. What make sociation possible at the biological level, makes socialism inevitable at the social level. Freedom of men in the (fictional) pre-contractual society was not a welcome experience. In the absence of social regulations on men's action, freedom turns out to be other's (and therefore also his own) burden. Tribes, communes, and similar other smaller social aggregates were the earliest social forms of the human efforts to reconcile freedom with law-bound necessity. Tribal socialism or primitive communism proved inadequate and restrictive: the emergence of private property, product of man's own labour, marked the end of tribalism and the beginning of the formation of larger human aggregates like nation and empire. This process of social evolution is also definable by such basic concepts as *diffusion* and *integration.*

Development means increased organization, Feudalism and capitalism are not as efficient as socialism in terms of organization. Mentally speaking, socialism involves self-transcendence and regulation of egoism, and without which clashing individual interests cannot be institutionally brought together and deployed efficiently for productive purposes. To make human life more comfortable, at least less unbearable on the earth, the socialist creed is necessary. Both Sri Aurobindo and Teilhard find an ontological of a cosmic justification of this necessity. Better and efficient evolution of life would be impossible without socialism. Having argued in favour of socialism up to a point, both Teilhard and Sri Aurobindo turn around and speak of its inadequacy. And the inadequacy is explicated in terms of centralism in the name of efficiency and integration. Intended integration of socialism ends up in a sort of "unintended" Statism, increased State power, seriously impinging on the freedom of the individuals and their choice of common identity. The law of Stateless freedom promised by the socialists like Marx proves ever receding and the demand for genuine freedom without compromising productive efficiency grows stronger. The theories of class-conflict, gradual pauperization of the proletariat and increasing polarization of the classes turn out to be suspect. "The old Marxist conflict between producers and exploiters becomes out-dated," and, two clashing spirits of the time are "the bourgeois spirit" which "simply wish to make the world a comfortable dwelling place" and "the toilers' spirit," "the spirit of changing the earth, the agents and elements of planetization."[10]

To Sri Aurobindo socialism means recognition of the inadequacy of egoism, the failure of the individual to recognize its own universality with other individuals, and the resulting concentration of power in some institutions like the State, investing it with more and more power. Though critical of socialism, Sri Aurobindo admits of its inevitability. He speaks of the "triumph of the socialistic idea or of its practice in whatever disguise—in all continents." He adds that "... after a cycle of violent struggle between the ideal of regimentation and the ideal of liberty the socialistic period of mankind might prove comparatively of brief duration like that of a monarchical absolutism in Europe."[11] One can hardly fail to notice here again the integration-diffusion rhythm of the evolution extended from the biological level to the sociological one.

Omega-Point or Life Divine

It is interesting to note that both Sri Aurobindo and Teilhard are confident that the socialistic period of man's social evolution would be brief and followed up by a period and a form of organization primarily characterized by freedom, flexibility and increased realization of the universality of human beings. Increasing interiorization of individual consciousness, according to them, releases higher powers of consciousness, enabling man thereby to establish larger sympathies with the fellow human beings and evolve more flexible and durable social organization necessary for the expression of their yet unrealized potentialities. Sri Aurobindo anticipates a *stateless society,* a free association of human beings. Like Marx, he is also an anarchist. While the former's materialistic anarchism is destined to get stuck up in a statism, 'a life-less machine,' Sri Aurobindo thinks that his own version of anarchism— spiritual anarchism— is in the very nature of things and expressive of a social necessity. Sri Aurobindo and Teilhard speak of the Kingdom of God on earth, which of course need not be taken in a literal sense. Once man discovers the law which unites him with all other beings and things, he becomes a citizen of a new world. The very process of evolution is, it seems, aiming at this yet unborn world.

II
POINTS OF DIFFERENCE BETWEEN SRI AUROBINDO AND TEILHARD

Pro-Naturalism of Teilhard and Anti-Naturalism of Sri Aurobindo

First, while Sri Aurobindo is an anti-naturalist, Teilhard gives a consistent impression that he is a pro-naturalist. Sri Aurobindo is bound to remind of Aristotle and Hegel, who, though recognized the reality of the natural world, were never tired of pointing out its lower level in relation to the mental level. Integralists and objective idealists have one commitment in common. They admit the reality of change and evolution and yet from the ultimate point of view find the same as indescribably transmuted. In other words, they want to reconcile temporality with eternity, multiplicity with unity, thereby derecognizing space-time as a principle of individuation.

More faithful to the scientific researcher, palaeontological findings in particular, Teilhard does not explicitly speak of teleological evolution. He wants to explain the evolution of the higher and the more complex forms of matter, life and mind in terms of new emergent properties. These properties were not eternally there. The world is not endowed with them right from the beginning. Teilhard fears that admission of such a position, the pre-existence of emergent properties, implies a sort of reductionism and/or essentialism, and belittles the importance of evolution itself. If psychology or noosphere is not allowed in its laws to employ descriptive terms which have not been used with approximately the same meanings in biology, the former discipline may be said to be reductive in relation to the latter. Judged by this notion of homogeneous reduction, the emergent properties of the mental world would be wiped out and de-recognized by the biological world, and those of the supramental world by the mental world, as we know of it today. If the proclaimed reduction of thermodynamics into mechanics is accepted as paradigm, and, following it, psychology is sought to be incorporated into biology, and biology in palaeontology, then the Aurobindo-Teilhard thesis of supramental evolution becomes an unwarranted research programme.[12]

No Ascent without Descent

Teilhard's pro-naturalism and pro-emergentism are not without difficulties. Sri Aurobindo finds it very difficult to explain the process of evolution without a complementary process of involution. In man's aspiration for the beyond Sri Aurobindo recognizes a power which is there in him right from the beginning. Teilhard's formulation of the point is somewhat different, understandably Christian in its implication. "Man's urge *Towards Something* ahead of him cannot achieve its full fruition except by combining with another and still more fundamental aspiration—one from above urging him towards Some One."[13]

Sri Aurobindo is explicit on this point. To him the descent of the spirit in the matter is a pre-condition of evolution. "In the descent into the material plane of which our natural life is a product, a lapse culminates in a total Inconscience out of which an involved Being and Consciousness have to emerge by a gradual evolution."[14] While Teilhard maintains that God from above and through man makes evolution possible and meaningful, a view which, to his mind, is perfectly scientific, Sri Aurobindo points out that unless God is deemed to have been doing double function, from above and also from below, the process of evolution remains half explained as the scientific account, or remains an enigma—as the Pedantic account turns out to be.

Life, Entropy, and Death

The controversy which has been going on in some form or other, from the middle of the last century, immediately after the First Law of Thermodynamics enunciated in the 1840s by Mayer, Helmholtz, Joule, and Colding, and the Second Law after 1850 by Clausius, Kelvin, and Boltzmann, needs to be studied without bias and prejudice. Some corollaries of the Second Law of Thermodynamics are said to be inconsistent with envisaged evolution of super-life from life. The Second Law is a statement of existence of stable equilibrium states and of special processes that connect these states to one another. The formulations of the Law given by Clausius, Planck, and Caratheodory are accepted as very authoritative. Equilibrium state means a stable state that does not change with time, while the system, the biosphere in this case, is isolated from all other systems in the environment. This state is not to be confused with the steady state, for the latter does not change with time even though the system is interacting with other systems. The value of any property of a system is expressible as a function of the values of the energy, number of elements, and constraints only. This may be said to be the first corollary of the second law. The second corollary is that irreversible process exist.

For a process is said to be reversible if the system and its environment can be resorted to their initial states. However, this condition, the restoration of the initial state and the environment of the system, is not satisfiable. And therefore, the corollary of irreversibility remains. The third corollary is that there cannot be any perpetual motion machine. If life is taken to be system in motion, it cannot move on forever. Because of this, the continuance of the life-movement involves work and the using up of energy. Used up energy brings about first equilibrium and then disintegration of the system. Implicit in this assertion is the assumption that the system in question is not interacting with the environment. To sustain the motion of the system, what is called for is continuous replenishment of the used up or diabatic energy. In the absence of interrelation between and interaction with, of the life-system with its environment, the system cannot be logically assumed to have the source for replenishing its lost energy. Unless it is assumed that life as an autonomous system has within itself an inexhaustible reservoir of energy, the perpetuity of the existing life-system or the evolution of super-life cannot be plausibly defended. Alternatively, it has to be assumed that lifesystem, although not autonomous and has no inexhaustible *internal reservoir* of energy, does receive not only its necessary replenishment, but also extra-supply necessary for the evolution of higher forms of life which are qualitatively richer than the existing ones.

One of the most important corollaries of the Laws of Thermodynamics is the principle of increase of entropy. The entropy of an isolated system may remain constant or may increase, but a decrease of entropy in such a system is ruled out. As we know, the life system, i.e. biosphere. is not an isolated system. It is related to, an evolute of, and interacting with, other systems of the universe. Accordingly, the life-process in nature, subject to the total entropy of all systems involved in the cosmic process, must either increase, or, if the process is reversible, remain constant. Unless we subscribe to the hypothesis of periodical or cyclical dissolution of the universe as a system, somewhat in the fashion of the Indian doctrine of *pralaya*, for instance, the cosmic process has to be taken as irreversible and the consequence of the principle of the increase of entropy has to be accepted. (Max Planck, *Treaties on Thermodynamics,* 3rd edition, 1927.)

The Earth, it has been conjectured, would undergo an almost incredible transformation in the next 1010 years. At the first stage sun would evolve into a much larger and hotter body, reaching almost the orbit of mercury, resulting, among other things, in the disappearance of the oceans, and due to excessive heat death of all forms of Life. The gas and dust in the galaxy would slowly disappear and form new stars. Milky way will become faint and dark. Bondi thinks that the answer to the question of creation will not be a matter of the past but left to the speculation of the metaphysician. It is a continuous process. "The average density of radiation must stay constant by virtue of the perfect cosmological principle and is replenished by the action of matter in stars which constantly generate fresh radiation. In this way the creation process, together with the expansion of the universe, prevents the approach of the heat death, the state of thermodynamic equilibrium in which no evolution can take place and in which passage of time has no significance." (Herman Bondi, *Cosmology*, Cambridge University Press, 1960, p. 144). Even if the universe as a whole can escape heat death, it is not clear how the biosphere can be credited with the same fate. The *steady-state* theorists, Bondi, Gold, and Fred Hoyle, cannot affirm that in new-born systems after the said heat death life will regain its lost forms, not to speak of the higher forms. Further it is to be noted that they rule out the very idea of evolution. The *evolutionary cosmologies* are certain that the dark and the relatively empty universe is doomed to greater darkness, emptiness and disintegration. Cosmos then reverts to the primeval chaos. Whether human life will rise again depends on the preservation of the identifying properties as it passes through an initial quantum cosmological state of almost infinite density. It is very unlikely that life would survive the cataclysmic inferno in which the stars and galaxies would be crushed by the dark, cold, sunless and long nocturnal span of time. Speaking scientifically, it is very unlikely that the vanished brilliance of the origin of the world, and with it new worlds of life and super-life, may once again reappear. This seems to be the upshot of the "evolutionary cosmologies" of James Jeans,[15] Georges Lemaitre,[16] and George Gamow.[17] Philosophers like Inge[18] are apparently persuaded of the correctness of this view.

The heat-death corollary of the Second Law of Thermodynamics has been strongly opposed, among others, by Oliver Lodge, Millikan, and the kinematic relativist, Milne. Milne thinks that this pessimistic view is not only "mistaken" but also "not an inevitable consequence of the Second Law of thermodynamics as applied to the universe as a whole." The crux of his argument is formulated to show that it is not logically possible to prove that some *parts* of the universe would not be able to escape heat death. And as regards the universe as a *whole* he asserts that it "has no age and no size" unless it is relativized to a particular observer (at a given place and time). In a religious vein he observes that it would be somewhat "irreverent" to think of an infinite God that He would be concerned with the biological evolution of only one "single planet," the earth, and would not exercise His godhead in infinite number of scenes in the infinity of galaxies.[19] Assuming it is possible, for a scientist, the question remains: What are the observable or other circumstances under which one could test the said hypothesis?

Both Sri Aurobindo and Teilhard, scientifically tempered as they are in different ways, want to settle their accounts with the implication of the scientific theory of entropy. In brief, the theory of entropy says that with the passage of time the earth loses more and more *usable energy,* becomes increasingly cold and is, therefore, doomed to die of cold. Either because of excessive heat at one point of cosmic evolution, or because of excessive cold at another point of the said evolution, the human species, it is apprehended, would disappear. If Sri Aurobindo and Teilhard have to defend their theory of the inevitable emergence of the Super-Life, of the Life Divine, or the Omega Point, they are logically called upon to meet this point of the natural scientist. Both recognize the seriousness of the question. Teilhard says, " . . . a grand Enigma presented by the phenomenon of man is not the question of knowing how life was kindled on earth, but of understanding how it might be extinguished on earth by entropy without being continued elsewhere." But he confidently affirms, "Having once become reflective it cannot acquiesce in its total disappearance without biologically contradicting itself."[20] Teilhard's optimism is understandable in terms of his religious conviction. but the question remains "how life defies cold death." A theological answer to the question to the effect that once the process of evolution has reached the high level of mind, it is inconceivable that it would suffer total destruction or disppearance, could carry conviction only with a person who believes in the teleological and cosmological proofs and in God as the architect and the executor of the scheme of evolution. But to say that or even to fall back on that position is to beg the whole question, and without answering the scientist's main query. To refer to the concept of irreversible time arrow in support of the claim that the evolution of the super-life is inevitable is not of much consequence either: Even those who like Eddington, Bergson and Bondi argue against the deterministic view of temporal becoming propounded by Grunbaum, Hugo Berg and others are not unanimous in interpreting it as meaning either (a) that the future events are without specificity of their time-reference; or (b) that the future time is quite unlike the determinable past and present. Inability to *know* in advance what forms the future of human life would assume does not give us a good title to assert that these are bound to be better and more potent than the recorded available ones.[21] Even if it is admitted what has been achieved through the passage of time, i.e., life, cannot be reversed, that does not mean either (a) that super-life is or (b) that life instead of tying reversed cannot be destroyed by lack of usable energy. Sri Aurobindo seems to be equally optimistic on this point and his optimism is based on spiritual and teleological premises. "Life is really not defeated by matter [for that matter by entropy]: it makes a compromise by using death for the continuance of life."[22]

All that Teilhard's and Sri Aurobindo's arguments against entropy add up to is this. Even if the Law of Entropy is accepted as valid, it does not mean the total destruction of life in the universe. This is what Milne has also strongly affirmed. It may mean, at the worst, that life may suffer a temporary setback here on this planet. With the nature of things being what it is, it is bound to appear here or eleswhere in a still more enriched form. Some philosophers of life-sciences have argued to the effect that the Law of Entropy is consistent with the *possibility* and survival of life even in the event of cold death of life on this planet. In other words, islands of life may survive in the ocean of destruction, due to heat or cold, in whichever way we put it.

Professionally, I am not competent to pronounce on the merits and implications of this view. But it seems to me that the argument is too slender to support the Aurobindo and Teilhard thesis regarding the advent of the super-life. From the mere possibility of the survival of life, it is difficult to be assured that the superlife is a must, a *necessity.* My point is simple: I do not deny this possibility. But the question remains: Is the possibility by itself a good enough ground to believe in and to develop such a comprehensive view? This question seems to me very serious because, for many people, it pertains to an area of life in which deepest sentiments and values are central. I, for one, do not like to suggest that one's religious beliefs and sentiments must be exclusively founded upon the theories of science, which, we know, change very fast. But snaping the relation altogether has its attending dangers.

A RECONSTRUCTION OF THE DOCTRINE OF CREATION AND DESTRUCTION

Having gone through the Indian doctrine of the circle of *srsti* (creation), *sthiti* (existence), and *pralaya* (dissolution), it often seems to me that the ancient Indian philosphers-scientists also addressed themselves to this basic question, the question of life and death and value. And they, in their own ways, tried to bring scientific views of the time and religious beliefs of the concerned people. By redeploying some of their arguments, drawn from different sources, mainly the *Samkhya-Yoga* and *Vaisesika* ones, one can meaningfully reconstruct the possibility of srsti (creation) after *pralaya* (dissolution). Strictly speaking, *pralaya* is a *quiescent* or tense state and not steady state. The effects of human action (*Karmaphala*) are said to be indestructible. According to the *Samkhya, pralaya* takes place when the actions of all human beings (*purusas*) collectively require a temporary suspension of all experience. The *Nyaya-Vaisesikas* think that the Supreme Intelligence brings about *pralaya* in order to give respite or rest to all living beings. All produced things disintegrate. The atoms and the souls with all their *dharma, adharma* and past impressions remain suspended in their own inanimate condition. *Adrsta* as accessory cause *(sahakari karana)* grounded in God's will brings about dissolution and sets in motion the forces of creation, disturbing the quiescent state. During a time of dissolution when everything returns to its original seminal position, the effects of human *karmas* are also preserved in a seminal form. The world before dissolution has a harmony of its own. The world after dissolution also retains this harmony in a seminal and inarticulate form. Nature preserves balance between all its elements, between the causes and effects of the same. This Law of Karma is cosmic in its scope and operation, encompassing the material, the vital and the mental actions and reactions. In the chapter on 'Rebirth and Other Worlds" in Life Divine, Sri Aurobindo argues to show that there is a Personality Principle of Legislator who retains a sort of harmony between the elements of the world to be reorganized and brought back into a life where it was left before the dissolution *(pralaya)*. This Personality Principle referred to by Sri Aurobindo might remind one of *Vaisesika* doctrine of *Adrsta*.

What breaks the quiescent state of equilibrium to make creation again possible? The question may be answered, following the *Samkhya* and without invoking the principles of God's will. It is enough to postulate that there is an orderliness in Nature, which is responsible both for temporary suspension and revival of the qualities, including activities of Nature. And this orderliness is related to the constancy of the total energy of the Universe. "The total energy remains the same while the world is constantly evolving; cause and effect are only more or less evolved forms of the same ultimate Energy. The sum of effects exists in the sum of causes in a potential form. The grouping or collocation alone changes, and this brings on the manifestation of the latent powers of the *gunas* (qualities), but without creation of anything new. What is called the (material) cause is only the power which is efficient in the production or rather the vehicle of power. This power is the unmanifested (or potential) form of the Energy set free *(udbhutavritti)* in the effect."[23]

One cannot deny the ingenuity of this argument in defence of the possibility of survival of life beyond death. But my old anxiety persists: Isn't its plausibility dependent upon the premises that there is a supreme intelligence which is powerful enough to harmonize the elements of the world, and retain the harmony in a potential form throughout the entire process of creation and dissolution. If this premise is withdrawn what substance is left in the argument and optimism of Teilhard and Sri Aurobindo? Following the classical *Samkhya,* which did not believe in God, one might say that harmony or unity of laws is there and that it has to be accepted as such. Obviously that is not the Aurobindo-Teilhard line of argument.

Teilhard says, "Entropy and life; backward and forward; are two complementary expressions of arrow of time. For the purpose of human action, Entropy... is without meaning."[24] "By its radial, i.e. internal nucleus the world finds its shape and its natural consistency in gravitating against the tide of disintegrational probability towards a diving integral focus of mind which draws it onward. Thus something in the cosmos escapes from entropy, and does so more and more."[25] It is by intense concentration and interiorization of consciousness, Teilhard claims, life goes upstream, defying death entailed by entropy of nature and, what is more, hominizing or spiritualizing death itself.

Somewhat in a similar way Sri Aurobindo wrote years before, "...such a survival of the self could only persist in the subtle body in the seminal form; being would still have to discard its physical form, pass to other worlds, and in its return put on a new body. The mental being is said to be capable of not only overcoming the physical causes of its decay and disruption but also capable of unfolding its secret spiritual divinity and the essential immortality of the Spirit in him."[26]

I often feel intrigued by the ingenuity, optimism and vision of Sri Aurobindo and Teilhard. I wish I could have the same experience and understanding and be happy in my intellectual life. But, unfortunately, the concepts of science, as I understand or misunderstand them, stand far as yet from the views of the fundamental point. Spinoza's *body-mind parallelism* has been carried too far in the "uneasy" *fact-value dualism* of Kant[27] and Einstein.[28] It is not enough to accept a unitarian framework, strong or weak, what, in addition, deserves recognition is the room of interaction between facts and values within it. True, it has not been possible to demonstrate the *strict correspondence* between or the *unity* of the laws underlying different spheres: but their *continuity* seems unmistakable, and, in a sense instructive.

The point to be borne in mind is this: religious sentiments, although do not owe their origin, enrichment, or impoverishment to the progress or regress of science, but when empirically established laws are shown to be in accord or at variance, with them they are *somehow* influenced—deepened, changed, disturbed, or intrigued. The religious sentiment itself can hardly be impervious to the feedback of experience. In this limited sense at least, religion has to respond to and settle its account with science. When the religious person, who expects to see the things in the world are law—following and *somehow* harmonious with his own existence and purpose, is reliably told of finds for himself that the way things are and would be in the world have nothing to do with his expectation and purpose or are inconsistent with the same, is bound to feel disturbed. If *per contra* the laws of the universe are found to be largely or perfectly in accord with his cherished values and purpose, he feels deeply moved. The third possibility, viz., that one might take everything that happens in the world with perfect equanimity, without being either moved or disturbed, is not logically ruled out. But, in that case, the problem we are discussing does not arise at all and therefore calls for no solution, either. This is a serious problem for those people, and their number and worth are considerable, who do recognize both religion and science as modes of experience and understanding. I know that it is possible to *dissolve* the whole problem by a painstaking analysis of the meanings of such words as *religion, science, life, death, existence, purpose*, etc. To my mind, this problem as a phenomenon is very much there, no matter how we describe it—theologically, psychologically, sociologically, or scientifically.

It goes to the credit of thinkers like Sri Aurobindo and Pierre Teilhard de Chardin to focus the problem and try to answer it in one particular way, may be without according due recognition to the objections raised by the scientists. It is for others to come forward and suggest the more promising lines of graphing it.

REFERENCES

1. Mario Bunge, *Treatise on Basic Philosophy, Vol. 3: Ontology I*, D. Reidel, Dordrecht-Holland, 1977, pp. 57-109.
2. Ernest Nagel, *Structure of Science*, RKP, London, 1971, pp. 368-72.

3. Pierre Teilhard de Chardin, *The Future of Man,* Fontana Books, London, 1964, p. 6I.
4. *Life Divine,* Sri Aurobindo Birth Centenary Library, Volume 19, Pondicherry, 1972, Book 2, Part 1. Chapter VII, "The Knowledge and the Ignorance", and Chapter X, "Knowledge by Identity and Separative Knowledge."
5. Pierre Teilhard de Chardin, *A Man's Place in Nature,* Fontana Books, London, 1971, pp. 18, 70, 71.
6. *Yajur Veda* 17. 67.
7. See above (4), Chapter XVIII, pp. 702-03.
8. Ibid., pp. 728, 907, 913.
9. Herbert Spencer, *First Principles,* 6th edition, para, 144. See also, pp. 280-82.
10 . See above (3), p. 144.
11. *Ideal of Human Unity,* Sri Aurobindo Birth Centenary Library, Vol. 15, Pondicherry, *1972, pp.* 390-91.
12. See above (2), Chapters 11-12.
13. See above (3), p. 302.
14. See above (4), pp. 663, 681, 684, 707, 912, 917.
15- James H. Jean, *The Astronomical Horizon, OUP,* London, 1945.
16. Georges Lcmaitre, *The Primeval Atom,* New York, 1950.
17. George Gamow, *The Creation of the Universe,* New York, 1952.
18. William Ralph Inge, *God and the Astronomers: The Warburton Lectures* 1931-1933, London, 1934.
19. E.A. Mime, *Modern Cosmology and the Christian Idea of God,* Oxford, 1952, Chapter X, pp. 151-52. See also, Herman Bondi, *Cosmology*, Cambridge University Press, 1960, p. 144.
20. See above (2), pp. 301-11.
21. Richard Gale, ed., *The Philosophy of Time,* Macmillan, London, 1968, sec. IV, "Human Time", pp. 293-386.
22. See above *(4), Vol.* 18, p. 233; *Vol. 19,* pp. 757-58, 804-14, 822.
23. P.C. Roy, *History of Hindu Chemistry,* Calcutta, 1901-1902, quoted from S.N. Dasgupta, *A History of Indian Philosophy, Vol. I,* Motilal Banarasi Dass, Delhi, 1975, Vol. I, p. 254.
24. See above (2), p. 81.
25. Pierre Teilhard de Chardin, *The Phenomenon of Man,* Fontana Books, London, 1971, p. 298. See also, pp. 56-57, 72, 318.
26. Pierre Duhem, *Aim and Structure of Physical Theory,* Princeton, 1954, pp. 287-91.
27. Kant, *Critique of Practical Reason.*
28. Albert Einstein, *Out of My Later Years,* Citadel Press, New Jersey, 1956, pp. 24-30.

Part II

Man-Environment Relations

Man-Environment Relations

Admittedly man can be studied in very many ways. But the two views which appear to me most revealing have been offered by the evolutionist and the ecologist. The evolutionist gives us a diachronic picture of man, tells us how species-man has evolved over millions of years, acquiring some properties and shedding some others. The ecologist, supplementing the diachronic picture of man given by the evolutionist, offers a synchronic picture of man in which the *contemporary* factors affecting man's nature and cultural life are identified and related. We should bear in mind that these two pictures are convergent and, in fact, one might say, two aspects of the one and the same picture. For, culturally speaking, past is present in man's life and consciousness or, as Grace and Collingwood put it in a seemingly paradoxical way, history is contemporaneous. Even after admitting the close relation between the said two pictures of man, we must note their difference which would provide added clarity to our understanding of man. A *series of successive pictures in the scale of time* gives us one sort of insight into the nature of man, whereas a *set of simultaneous pictures* in the *frame of* time gives us another sort of understanding of the factors, natural as well as cultural, dialectically shaping what man is. Further, socialist-humanists point out species-man, though now better *understood,* is yet to be achieved. I have discussed this issue in *History, Society and Polity* (Ch. 11).

In the essays included in Part II, I halve tried to show, broadly speaking, two things: one, man is informed of his environment, both natural and cultural, but he is conscious of only 'a part of this information system, and, two, the conscious part of his information is expressed, articulately or inarticulately, in his thought and action which, in turn, influence his environment. This dialectic goes on. And within this general dialectic we can discern, on analysis, some local or particular areas of dialectic, both at the level of nature and also at that of the cultural life of man.

The gradual relation and distinction between nature and culture deserves close attention and study. The sharp line of distinction that is often drawn these days between the sciences of nature and those of culture, or between science and philosophy, was not there before, not even in the last century, and that would be evident both from contemporary books on history of science and history of philosophy. Some sort of distinction "corresponding" to what we now call experimental knowledge and speculative knowledge, it seems, was there in the past, both in the primitive and the medieval society. The said correspondence between the forms of classified information of today's and those of yesterday's presuppose certain other general forms of classification and the availability of rules of translation between them. Increasing information-explosion necessitated classification and reclassifi-cation, transforming old or existing systems of reference and forming new ones. Sometimes this is conveyed by such terms as "de-totalization", "re-totalization" (Sartre), "paradigm change" (Kuhn), and "change in research-programme" (Lakatos).

The traditional theories on the nature and origin of knowledge, rationalism and empiricism, for example, appear somewhat narrow and unrelated to the relevant factors of physical nature, biology and sociology. One may attribute this unfortunate development again to what is called information-explosion. In the past, the principle of division of intellectual labour was not pushed to an absurd extent. In some parts of the world, especially in some areas of learning, biopsychology, biolinguistics, information theory, etc., the "interdisciplinary" approach is staging a comeback and trying to undo, at least partially, the damage done to the search for knowledge by the overzealous specialists. Fruitful search for knowledge continuously needs informed interaction between the specialists and the generalists and also between the specialists themselves.

The questions that initially bother one is what are the cognitive, linguistic and cultural universals found necessary for making interdisciplinary, inter-cultural or inter-societal communication possible and meaningful. In essay 6, I have tried to show the close relation between ecology and epistemology, using the rudiments of information theory. The cognitive system, the knowing body-mind complex, is informed of many things it needs and yet not conscious of the same. We, as knower, are being fed by our environment and the presupposition is that we, human beings, are endowed with the psychophysical competence of receiving, organizing and extrapolating the informations, the encoded messages, given by the environment. The classical rationalist was wrong in thinking that the knower is a "transcendental self" not quite involved in what happens in the body or the sensuous mind engaged in knowing, in possessing information. It is not always adequately realized that the knower is an embodied human being and bears the legacy of having body in the acts of knowing. Man is after all an anthropoid, an animal, and subject to the laws of nature. To say this is not to degrade man and level him down to the levels of animals who are not capable of forming, relating and using symbols. I have tried to indicate the continuity between the human and the sub-human species. On the contrary, inadequate realization of this important fact often feeds romanticism of reason and disturbs the balance between man and nature.

Having clarified this point I turn my attention, in essays 7, 8 and 9, to some theoretical issues, illustrated by concrete cases, of history of science. I have taken the terms "science" in its wider connotation. Today's laboratory sciences are not all that I mean by science. As we know, the scopes of *science* and *philosophy* are being defined and redefined. The subjects falling under them, and also other growing disciplines, are being allocated and re-allocated. The fact that, in spite of all these changes, the historian of today can know the scientific ideas of the yesterdays, demarcating them from empty speculation and tracing their relation to good metaphysics, is indeed very revealing. Equally revealing is the fact that the anthropologist, affiliated to and nourished by one culture, can understand the ideas and actions of the people of alien cultures. What is revealed is this: physical (spatio-temporal) distance and cultural difference, though they must not be berated, can be substantially overcome. At this point the admitted role of linguistic and cultural universals has been overrated by some such thinkers as Levi-Strauss and Chomsky. In their understandable eagerness to explain *many* in terms of *few,* to construct theories with strong explanatory powers, they appear to have failed, in part, to recognize adequately the significance of the *diversity* and the *evolutionary* character of the phenomena to be explained. There are indeed genuine differences in understanding foreign languages and alien cultures. It is not for nothing that the philologist and the anthropologist arc obliged to work so hard and go into factual details without being lost, as the positivist often is, into the same. In order to avoid the error of the positivist, the structuralist setting his feet in the direction need not go to the opposite extreme or error. Explanatory universals as well as factual particulars have to be taken seriously into account. The dialectic between the two has to be gone into. In their different ways Popper, Quine, Sartre and socio-linguists like Bernstein and Halliday have tried to emphasize this (not a very new) point. But Levi-Strauss's negative reaction to Sartre's criticism and Chomsky's to Bernstein's suggest that the importance of the point is yet to be adequately recognized.

In essays 7, 8 and 9, I have tried to explain constructively, without entering into extensive polemics, how intercultural understanding and communication are in fact made possible. Both Kant and Einstein felt deeply impressed by the fact that the human mind can at all comprehend the universe both in its vastness and minuteness. Rightly understood, this is not an ordinary fact as it might appear to a layman. If the comprehensibility of the natural universe proves so intriguing, one can well imagine how intriguing would prove the comprehension of a cultural universe marked by even more diversity and inconstancy. Kant's proposed answer to the question, the human mind can comprehend nature because it is endowed with the *categorical competence* to create it, though insightful and promising, has been rejected, I think rightly, by Einstein and Popper on the ground that the categories are *not fixed.* Neither, competence nor performance of the human explainer is categorical in the Kantian or the Cartesian sense. The ways in which human beings, natural and social scientists, explain natural and cultural phenomena are partly to be found in our body-mind complex and partly in the phenomena themselves. These two parts cannot be functionally isolated but its denial can be disproved. The explanatory ways or theories, as a part of the changing world, do change. Simply because they explain change they are not themselves above change. True, they change at a different level and, relative to the phenomena to be explained, slowly.

That scientific theories change is a generally accepted view. But that philosophical theories influence the changing career of science and that scientific theories, in turn, give rise to new philosophical theories are not widely appreciated views. In fact the views expressed in 7, 8 and 9 are related and complementary. In them, among other things, I have tried to show that the diachronic or historical method and the synchronic or structural method are not antithetical as it is often sought to establish. These two methods, to my mind, are two aspects of one and the same dialectical method. The dialectic between facts and theories bring out the increasing and changing intelligibility of facts and the tested modification of theories. We, humans, can understand nature because (a) we ourselves are part of nature, (b) laws are there in nature, and (c) we can reflect over our partnership with nature and the laws which dynamically influence it.

6

Dialectics of Ecology and Epistemology

> We must trace the movement that produces life out of matter, man out of primitive forms of life and social history out of the first human communities. —Jean-Paul Sartre[1]

Long before the advent of the discipline now called biophysics the ancient thinkers of India and Greece spoke of the exchange of powers between the living world and the non-living physical world and were logically obliged to speculate the existence of like powers in both the world and, thus, explain the evolution of the biosphere on the earth. In the *Principia* Newton wrote of "a certain most subtle spirit which pervades and lies hid in all gross bodies," and that "all sensation is excited, and the members of animal bodies move at the command of the will, namely, by the vibrations of this spirit, mutually propagated along the solid filaments of the nerves, from the outward organs of sense to the brain, and from the brain into the muscle." He was looking for a scientific hypothesis to explain the relation and the environment. The same problem at a different, i.e. at a more speculative, level had engaged the attention of the atomists like Democritus and the fluxists like Heracleitus.[2] While the former thought that life results from attraction between and combination of like atoms, the latter having realized the uncomfortable implication of this view, viz. unlike atoms oppose each other, and having observed that apparently unlike objects do in fact attract each other, maintained that all changes—attraction, combination and collision—are only ceaseless exchange according to law. It seems the fluxist had a vague notion of inertial motior and more promising idea of dialectic-conflict without contradiction. At one stage it was widely believed that the atomistic approach, improved by Dalton's chemistry, would provide a satisfactory framework for structural or metabolic aspect of biology, and that the fluxist approach, favoured by Bergson and Whitehead, would give a better account of the developmental aspect of biology. But the later researches have shown that even in metabolic activity for structural preservation non-repetitive changes in instructions and informations received by individual organism both from without and within are necessary. Received information are not only organically preserved but also hereditorily transmitted with suitable modification, selective and rejective-suitable for the adaptive purposes.[3] The supposed distinction between metabolism and development is untenable and this is evident in the viruses and bacteria: metabolic activity of these organisms consist of the development of new organisms.

Corresponding to matter-motion or particle-wave dualism in physics, one could speak of plasma-soma (Weismann) or its modern equivalent DNA (deoxyribonucleic acid), PFC (protein, fats, carbohydrates, etc.) dualism in biology. In the light of recent researches and interpretations, the latter dualism somewhat like the former one is being rejected by most of the experts. Because the instructions contained in plasma/DNA, it has been pointed out, do largely influence the functioning of soma/PFC synthesis.[4] In fact the biological process is marked by creative interplay between the genetic instructions and the mechanism by which those instructions are carried out. Like the physical process the biological one also is increasingly appearing as creative continuum.[5] Using metaphysical language one might say that being and becoming are two aspects of one and the same process and not antithetical. Speaking in a less metaphysical vein the Marxist claims to have succeeded in interpreting the whole process dialectically, point out the inner connection between necessity and chance and the creative interplay between heredity and adaptation.[6]

As evident from their writings and letters of the 1860's and 1870's Marx and Engels were deeply impressed by Darwin's theory of evolution, particularly its concept of natural selection.[7] Against the backdrop of the Malthusian theory of population and their critique of the capitalistic mode of production and exploitation, this is very understandable. But they pointed out that in the concept of natural selection Darwin conflated two absolutely separate issues: (a) selection by the pressure of over-population, and (b) selection by greater capacity of adaptation to altered circumstances. In the absence of cautious formulation "a" might mean not only survival of the fittest but also of the weakest at times, and "b" not only adaptation and progress but also regression and conservation. In that case evolution becomes indistinguishable from the dance of chance and ceases to have any dialectical *rationale* in it. The true law of evolution seems to be this: regression, if and when takes place, is at the instruction of the circumstances and which is conserved and used to make progress possible in one direction, excluding the possibility of evolution in many other directions. The aspects of evolution, progress and regress, are the result of dialectically imparted environmental instruction. Neither the struggle of opposites in nature is lawless nor the struggle for existence in society utterly senseless.[8] The Darwinian theory of natural selection unless correctly interpreted, the Marxist fears, would be a pernicious mixture of the Hobbesian theory of war of each against all and the Malthusian theory of population and competition as transferred from society to organic nature.

According to the Marxist, everything from nature to culture, including thought, is pervaded by dialectics, by the ceaseless conflict of the opposites and their exchange, or passage into higher forms. Polarity is exhibited by magnetism, electricity and chemical processes. This polarity is said to be best accountable by dialectical materialism and not mechanical materialism. The *objective* dialectics is operative also in organic life and its evolution, both marked by a polarization of the living protein material and the ceaseless conflict between heredity and adaptation. Because of the ceaseless conflict and exchange of properties between the opposites, the Marxist refuses to recognize any one of them either as absolutely positive or as absolutely negative.[9] For instance, heredity at times appears as positive since it conserves and transmits the acquired properties of one generation into another, and adaptation as negative since it seems to destroy at least partially, the inherited properties. But one could plausibly look at the whole thing other way round since adaptation enables the living being to survive and procreate while heredity plays the role of passivity, resistance and negative activity. But at every nodal point of organic evolution and critical epoch of human history, it seems that adaptation practically plays the relatively dominant negative role, partially negating the forces of forward movement. The Marxist is cautious enough to maintain that it is not the only possible way of looking at the matter. The question may be raised whether the relativity of position and negation is a matter of *interpretation* or *description* of the physical and chemical processes and of the biological evolution.

The Marxist's response to the question is clear from his claim that all pervading dialectics is *objective*. Since the composition within and relations between the basic units of matter and life dialectically and ceaselessly change, the objective description itself, to be truthful, has to undergo change. Otherwise in course of time it turns out to be unfair to the facts of the world. Besides, the sort of monism the dialectical materialist propounds, being quite different from the monism "oceanic unity" of the "objective" idealist, is consistent with the separate existence of different basic entities, matters-in-motion, and leaves room for their dialectical interaction, accounting for their internal and external exchange of properties and consequential change over time. It may be added here that similar views were expressed at that time by some other non-Marxist scientists as well.

While Marx and Engels were doing epoch-making works in social sciences in the 1860s and 1870s, Darwin, Galton and Mendel were engaged in making significant and in some respects complementary contributions to *genetics,* although the term was not in scientific use before 1906. Mendel thought that genes remain pure and separate at germ-cell formation, i.e. not contaminated by the parent organism and its environment, that they combine randomly at fertilization, and replicate themselves identically unless some random mutations of the genes become the only source of heritable variation and that their new characters appear suddenly without any previous selection. The Darwinians pointed out that mutations not gradually instructed by environment were incapable of explaining the origin and improvement of adaptation. Soviet biologists, Michurin and Lysenko, taking cues from Marx (letter to Engels of August 7, 1866) and Engels have attacked the Mendel Weismann doctrine, criticizing the inheritance of acquired characters.[10] To what extent their anti-Mendelism was scientifically sustainable was a matter of prolonged dispute, but its ideological motivation was clear: for the Mendelian genetics was incompatible with the Marxist idea of shaping the future communist man, "the total man". The political fall of Lysenko with Kruschev does not necessarily mean the full vindication of Mendelism in all its aspects. While the importance of Mendel's discoveries continue to be regarded as outstanding in the history of genetics, its shortcoming, ideology apart, can hardly be ignored. Mendel seems to have an inadequate appreciation of the objective dialectics between heredity and environment. Evolution is best definable as change in the genetic composition of a species through time and it progresses on two levels, the production of variation and the selection of those gene combinations which are found to be most fit in a particular environment.[11]

Fitness of a population means basically its competence to cope with its environment, both natural and social, at a particular time. The factors that influence the course and direction of its evolution are numerous and vary in their effectivity, both directly and cybernetically : some more noteworthy of them are disease and predation; conflict and migration; behaviour and disposition; competition for breeding space and mates; competition for living space and food.[12] The wider ecosystem and the immediate ecological, including the social conditions, and some yet-not recognized factors—all have their say in and take their toll on our more and more freedorn-seeking, organization-building and yet entropy-threatened destiny.[13] How man's freedom-seeking dispositions and origination-building activities prove rewarding and, even without manifest reward, useful, is a fascinating dialectical process marked by occasional integrations of man and environment and occasional separation of the former from the latter, but all the time both bear the selective imprints, not necessarily perceptible, of information-exchange.

What happens at the biological level is repeated anew, defined by qualitative enrichment, at the psychological level as we gather from the scientific studies in the processes and forms of learning.[14] It is true that there are differences in neural structures of lower, and higher organisms as evident from the difference in their ability to make abstractions, form concepts, and make use of them in coping with the problems they encounter, but it seems certain that there are some characters controlling behaviour which are common to all of them. The human ways of learning and the non-human ways of the same are not entirely different and do structurally overlap in some very important respects. Habituation, conditioning, imitation, trial-and-error, etc. are often said to be peculiarly non-human or animal ways of learning. And, in contrast, insight, association, meaning storage, tool-using and constructing, reinforcement and conceptualization, etc. are claimed to be the main characteristics of the human learning process. Careful analysis, however, shows habituation, for example, involves capacity to abridge and preserve the results of experience, a capacity which may or may not be recognized as concept-forming or meaning-storing. From the other end one can point out that insight or conceptualization is not as spontaneous, "rootless" or a priori as philosophical rationalists always try to make out.

Even what is ordinarily known as conditioned reflex learning is, on analysis, found to be much more complex than hitherto believed. In the light of the advanced research works in the field it is now becoming increasingly apparent that the "conditioned reflex" in higher animals is not primarily "pushed" by stimulus from behind as cause but "pulled" by anticipation or expectation. Pavlov failed to see clearly that a dog salivates not only in response to a buzzer but also in studied expectation of food.

The trial-and-error method of learning exhibits even more articulately how error-detecting experiences influence the successive subsequent trials with increasing appropriateness, appropriate for the purpose of getting to the desired end or reward. But one should be cautious and bear in mind that animals like rat and kitten cannot directly perceive the relation between their trial-actions and the desired reward. What happens in their case is like this: even the unsuccessful or erroneous trial-actions leave in them some or other trace, reinforcement-instruction, enabling them to select a less inappropriate response from among a variety of trial-actions objectively related to the desired reward. With the progress of the process, it seems, an increasingly closer relation is established between the stimulus or situation and the appropriate response for getting the reinforcement or reward. Each partial reward appears to be preceded by an anticipation, however vague it might be, of the requisite response.

It is often mistakenly assumed that animal learning is necessarily reward-oriented or reinforcement-seeking. Unless the concepts of reward and reinforcement are suitably redefined taking duly into account their latent characters, this assumption would clash with many experimental findings.[15] Reward need not be always in the form of food or some such gross object. Whatever facilitates an animal's or organism's capacity to adapt itself better with its environment or is of use to it in the matters of identification and elimination of erroneous or inhibitive responses and thereby, positively speaking, in "choosing" the right responses, is to be recognized as reward. Right sort of orientation or augmentation of an organism's sensory-motor activities, enabling it, may be marginally, to get closer to its gross object of desire or need, certainly does act as a positive reinforcement. One may rightly claim that whatever helps the process of learning, to learn is in an important sense, itself a countable reward. Even in the non-human levels of animals, birds and insects learning proves often, if not always, very latent. If this argument is accepted, I do not see why it should not be, the line of demarcation drawn between the human ways of learning and the non-human ones becomes more and more blurred and untenable. By extending the argument one can also plausibly affirm that neither the insight method of learning is peculiarly human nor the trial-and error one peculiarly animal. For the ingredients of latent learning are to be found in varying degrees in the both. To put the whole matter more dramatically, not mistakenly, I believe, one can easily cite experiments showing that chimpanzees and rats, for example, do learn with the help of their insight, while their human counterparts are obliged to learn through trial-and-error as evident from somewhat unpredictable progressive-regressive shifts in history. History of science abounds with supporting examples as pointed out by Lakatos and Feyerabend. Soviet genetics and Nazi Biology are two among the most recent examples of regressive shift.

Because of their less organized bodies, in the absence of some organs or their reduced specific qualities, certain organisms, insects and animals, are not capable of learning from the environment what men are.[16] Open to the same objective stimuli (of heat, colour and food, for example) while men, because of their specifically developed sensory, perceptual and motor mechanisms, can grasp, even anticipate them with relative ease; animals fail, more so in strange contexts or new environments. The capacity to abstract, generalize and correct, marks human learning: therefore, when man is placed in environments other than the one(s) with which he is familiar, he may find difficulties but is not likely to be over whelmed and destroyed. In contrast, fishes of one hemisphere do not survive in another. This is true in cases of various species of birds as well. In brief, the human world is big, but the animal one limited. The limitations of learning has to be studied from both ends, the end of the respondent, animal or human, and its neurophysiological organization and also from that of the environment, the source of stimuli and reinforcers. And this study is required to be dialectical as I have already indicated before, response and stimulus, respondent and environment, are in a process of exchange, and work interdependently in a sort of field. This recognition of the mutuality of respondent environment relation *in a field* is, by implication, a corrective both of crass empiricism and apriorism in the area of learning. That men and animals are endowed, though not identically, with some neurophysiologically founded learning capacities has to be admitted without hesitation.[17] It has also to be admitted that those capacities are neither innate nor work unilaterally and uniformly without being informed and influenced by perceptual-motor mechanisms of the concerned organism—human or animal.

REFERENCES

1. Jean-Paul Sartre, *Critique of Dialectical Reason,* 1976.
2. Ernest Hutten, *The Origins of Science,* 1962.
3. Rene Dubos, *Man Adapting,* 1965.
4. J . C. Eccles, *The Physiology and Nerve Cells,* 1957.
5. Joseph Needham (ed.) *The Chemistry of Life,* 1970.
6. Frederick Engels, *Dialectics of Nature,* 1954.
7. Charles Darwin, *The Expression of the Emotions by Man and Animals,* New edn., 1965.
8. Jean-Paul Sartre, *op. cit.*
9. Frederick Engels, *op. cit.*
10. A.H. Sturtevant, *A History of Genetics,* 1965.
11. Julian Huxley, *East and West: Lysenko and World Science,* 1949.

12. Sol Tax (ed.), *The Evolution of Life: Its Origin, History and Future,* 1960.
13. Theodosius Dobzhansky, *Mankind Evolving,* 1962.
14. E. R. Hilgard and G . H. Bower (eds.), *Theories of Learning,* 3rd edn., 1966.
15. Wesley E. Lanyon and W.N. Tavolga (eds.), *Animal Sounds and Communication,* 1960.
16. R. A. Hinde, *Animal Behaviour: A Synthesis of Ethology and Comparative Psychology,* 1970.
17. Karl R. Popper, *Objective Knowledge: An Evolutionary Approach,* 1972.

7

Towards a Philosophical History of Science :

Some Preliminaries

Let us think of a society called Primitive Society (in short PS) at a time T1 with a number of wise men called Magicians (in short M) in it. Let us think of another society called Medieval Society (in short MS) at a time T2 with a number of wise men called Philosophers (in short P) in it. Let us think of still another society called Contemporary Society (in short CS) at a time T3 with a number of wise men called Scientists (in short S) in it. Let us take PS, MS and CS as *ideal types of* historical societies, i. e. not description of *this* or *that* actual society. T1, T2 and T3, the time-addresses of PS, MS and CS, are also *ideally* sliced segments of history, i.e. not photo-mnemonic reproduction of *this* or *that* particular time. PS (TI), MS (T2) and CS (T3) are from my point of view, only conceptually successive; and I need not suppose that they are successive in the objective time-order.

Ideally sliced segments of history can be simultaneously pictured before the eye of the mind and studied as separate parts of one single unrolled carpet of time, or, should I say, slightly curved canvas of space. Simultaneity and histority of PS (TI), MS (T2) and CS (T3) are not intrinsically incompatible concepts. The "same" phenomena, in this case societies, can be studied from two different but complementary points of view, historical and sociological. History may be sociologized and sociology historized. Different and separate societies, e. g. PS (T1), MS (T2) and CS (T3), may also be looked at as ordered in the scale of time. When I see segments as segments I have already in a sense spatialized time.

And it is no wonder that one raises one's eyebrow when one hears for the first time the theoretical term *contemporàneity* of history. To break the spell of wonder what one is to be reminded is that it is impossible for one's understanding to escape its basic affiliation to its cultural environment which includes, among other things, theories, beliefs, norms and forms of action. For example, when I am trying to understand diachronically, i.e. historically, or synchronically, i.e. sociologically, science as a particular set of activities, or a corpus of beliefs, or both, I and others as well must reflectively bear in mind that my understanding itself is culturally affiliated to one or other CS (T3). Affiliated as I am to CS (T3), in certain respects I may be enjoying more advantages than you of MS (T2) or he of PS (T1) in the matter of understanding the ideas of the wise men, S, i.e. scientists, but, with my understanding embedded in CS (T3), it proves often very difficult for me to turn my understanding on the presuppositions, assumptions, shared beliefs, etc. which are almost constitutive of CS (T3). Moreover, culturally at a distance from, or being somewhat alien to, MS (T2) and PS (T1), I cannot easily get into the structures of the ideas of the wisemen of MS (T2), i.e. P, and those of PS (TI), i.e. M. The structures of the activities and ideas of S, P and M are not only conceptually successive but also partially overlapping and mutually inclusive, and, therefore, cross-cultural identification, cognition and communication are possible.

Let us now think of a society called Future Society (in short FS) at a time T4 with a number of wisemen called Super-Scientists (in short SS) in it. It is also an ideal type society which can only be structurally (and not specifically) described. Our forecast about the activities, achievements and beliefs of the laymen, especially of the wisemen or SS, of FS (T4) is bound to have a ring of fictionality around it for my ear used to the tune of S of CS (T3) for years.

Fact/fiction dualism is primarily, but not entirely, intellectual and contextual. The net of fiction is cast to catch the elusive but supposedly existing fishes of facts. The cast of the layman is wide and that of the wiseman narrow, for the former knows only vaguely what facts he is after, while the latter does it pretty precisely. The line of demarcation drawn between facts and fictions or that between science, on the one hand, and non-science, comprising myths, metaphysics, and other forms of speculation with thin content, on the other, and the criteria used for the purpose of drawing the line are somewhat peculiar to the concerned society or even a sub-society within it. The nature of the problems under consideration suggests the line and criteria of demarcation. As already stated, the wiseman/layman distinction should be understood as a culture-relative way.

I may be accused of indulging in an extreme form of relativism and my proposed theory of the history of science and philosophy criticized as a purely contextualist and therefore ad hoc one. I plead not guilty to the charges. I would submit that to avoid the more serious charges of anthropomorphism, cultural hegemonism, ethnocentricism, or even absolutism, one is perhaps well-advised to be a critical contextualist. When I say "my understanding of other societies is bound to be shaped by the structure and capacities of my understanding itself and which in their turn are influenced by my culture", I am giving a mere reflective description of a complex situation and trying to avoid the fallacy of the absolutist-rationalist whose unilateral "cognitive" verdict on other cultures seems to suggest' that he has a privileged access to the meanings and aims of the activities and beliefs of other persons, e.g. M of PS (Tl), P of MS (T2), S of CS (T3) and SS of FS (T4). My question is : who am I, or for that the absolutist is to say that they do not know what they mean when they say, do and believe something, or, what is worse, that they are not logically equipped, for the purpose. It is better we recall that not only ethics and economics but also geometry and logic have behind them long historical career. We should also remember that culture-oriented character of one's understanding is no hindrance to one's understanding of others' activities and beliefs, i. e. magic, philosophy or science. It is evident from numerous perceptive histories of science written from articulate or inarticulate sociological or anthrological points of view. Of course there is always a difference between a good history of science and bad one, whatever might be the author's point of view. All things being equal, a historian who is reflectively and critically conscious of the sociological linguistic and other cultural factors which influence his understanding is expected to give more correct accounts of others' achievements, scientific or otherwise, than the accounts given by those who are not so conscious. Context-invariant and theory-neutral history of pure observation language is a chimera.

Every society has its own models of layman, wiseman, physician priest, philosopher, scientist. But the words do not mean the same things everywhere, PS (TI) or MS (T2) or CS (T3). Even within our own society, these words are used to mean different things in different contexts. Sometimes the layman thinks and acts as the wiseman does; the opposite thing also happens. Sometimes the philosopher performs the duties of the priest and also those of the scientist. What is very important to bear in mind is the precise time-context we are talking of. In PS (TI) and MS (T2), e. g., the roles of the wiseman, the physician, the priest, the philosopher and the scientist had not been sharply defined, rather they all had been practically rolled into one. Compared to what one sees in the CS (T3) the principles of division of labour were much less carried out in PS (TI) and MS (T2) and, given the present trend, are likely to be even more practised in FS (T4). The authors of the Vedas had in them various qualities admirably harmonized, via. poetic-artistic talents, scientific insight, philosophical vision and power of thought. The early Greek cosmologists—Thales, Anaximander and Anaximenes—were also both scientists and philosophers. The point has been persuasively argued, among others, by Ueberweg. Windleband and B. N. Seal. "If by science we understand that independent and self-conscious work of intelligence which seeks knowledge methodically for its own sake, then it is among the Greeks... of the sixth century B.C., that we first find such a science—aside from some tendencies among the peoples of the Orient, those of China and India particularly, only recently disclosed. The great civilized peoples of earlier antiquity were not, indeed, either in an abundance of information on single subjects, or in general views of the universe; but the former was gained in connection with practical needs, and the latter grew out of mythical fancy, so they remained under the control, partly of daily need, partly of religious poetry ..."[1]

It is instructive to bear in mind that the great names in Hindu medicine, *Susruta* (fifth century B. C.) and *Caraka* (second century B.C.), and those in Graeco-Roman world, Hippocrates, a contemporary of Susruta, and Galen, a contemporary of Caraka, mastered several other branches of knowledge than the one with which their fame remains most intimately associated. All of them had thorough grounding in the First Principles of Knowledge, or, what we call philosophy these days. Leonardo da Vinci (1452-1519) and Leibniz (1646-1714) were versatile genius of the very rare type; the former was not only an extraordinary painter, sculptor architect and engineer but also a gifted scientist and a philosopher, the latter left his marks on such diverse fields as logic, mathematics, mechanics, geology, jurisprudence, history, linguistics and theology. Neither in the early ages nor in the middle ages the principle of the division of labour was in the field of knowledge as we see it practised today. Every age, as I said, has its own intuitive, i. e. undefined notions of the layman, the wiseman, the physician, the engineer and so on and operates with the same. Of course the notions change slowly over the centuries, over the decades; but at times the notions change fast and almost in a revolutionary manner. I say "almost" because the revolution itself has its own seminal history and, once it is brought about by the genius of an Archimedes, or Arya Bhatta, or of a Copernicus or an Einstein, it does not take at least the informed by stupefying surprise.

Another point which is very important and of which we can be easily conscious on reflection is this. We tend to accept most of the parts of our cultural heritage, except in respect of those about which we can claim to have first hand or professional acquaintance, almost as a matter of course, i.e. without serious questions or criticism. As I happen to be a professional philosopher, my questioning attitude and critical outlook are almost entirely confined to currently influential philosophical theories and arguments. In so far as other areas of knowledge, are concerned, e.g. medicine, metereology, economics, engineering, physics and, especially their higher reaches and latest findings, we generally accept, rely and act on the specialists' views. To us, the outsiders and the laymen, specialists' views are an integral part of culture. Unless these views prove inconsistent with or problematic to our otherwise conventionally accepted culture, we do not turn our critical queries on them.

Ordinarily cone is not critically disposed to the whole of one's culture, whether that culture is PS (TI) or MS (T2) or CS (T3), or whether that one is a layman or a wiseman or a specialist in the concerned context. One tends to believe and accept all non-problematic elements or aspects of one's culture as almost inseparable parts of one's being. Language and all that is preserved in it, customs, myths, conventions, etc. not only environ us but also penetrate deep into our personality structure, obviously not in a uniform or monotonous way. The problematic features of a culture can hardly be identified and communicated in a person-invariant, i.e., purely objective, manner. The symptoms of a person's bodily behaviour which appear to me absolutely normal may immediately disturb an expert diagnostician who can read in them a danger signal of a grave illness. Problem is an objectively disturbed area of one's conceptual network and enables one to truly think of, and successfully act in the world one lives in or tries to understand. Generally speaking, we take only some practical problems, war, strike, epidemic diseases, energy crisis and inflation, e.g., as universal or at least near universal in their scope. In respect of theoretical problems the scope becomes universal only in those rare cases when the proposed solutions result in revolutionary changes in the prevalent world views—the sort of revolutionary changes associated with the names of Marx, Freud and Einstein in CS (T3). One is advised to bear in mind that there are some theory-intoxicated men who take their theoretical problems with more ardour and concern which characterize the common man's practical problems with the runaway rise in the consumer's price-index, transport strike or energy crisis. Theory-intoxication may be of a different sort resulting in distortion beyond recognition of other cultures in one's own culture.

That most of us accept our culture, at any rate most of its elements, rather uncritically, think and act conventionally, has been very persuasively analysed, among others, by Hayek in a somewhat different, although related context. In the course of giving his view on the nature of the subject matter of social studies he says, "A medicine or a cosmetic, e.g., for the purposes of social study, are not what cures an ailment or improves a person's look, but what . (the concerned) people think will have that effect."[2] To say this is not to suggest that the subject matters of all social studies are equally and irremediably subjective or remain identically so over a long period of time; and still less does it mean any offence to a realistic world-view, i.e., that the World (in a sense) is there independently of how it is thought of, e. g., by M of PS (TI), P of MS (T2) or S of CS (T3).

A word of caution. The ontological question: "Whether there is World independently of its being perceived or conceived by a man or a group of men?" is logically different from the epistemological question: "How that independent World is perceived or conceived?" True, to make the answer to the ontological question clear and intelligible, it has to be brought closer to, and perhaps formulated, to a great extent, in terms of the answer to the epistemological one. Conceptual inseparability of the answers to the questions does not logically oblige one to believe that the imports of the questions themselves are identical. Laying too much of emphasis either on the slogan of *esse est percipi* or on that of *esse est concipi* we are avoidably caught in an erratic and almost endless swing between the two extreme forms of constructivism. I am not opposed to constructivism in its every form. On the contrary, some persistent features of *every* cultural life, viz., interpersonal and intercultural communication of ideas and experiences, cross cultural identification of objects, and so on, are indicative of the *general* truth of methodological constructivism. Methodological constructivism is not inconsistent with ontological realism.[3] But, unfortunately often the one-sided advocacy for the case of constructivism tends to make us blind to the fact that *general* character is ontologically warranted and sustainable. And this would be evident both from the critical reflections on the findings of the working anthropologists and the conclusions of the historians of sciences.

The concepts of relativism, contextualism and constructivism are closely interrelated. Theories and practices, science and technology of a society are best understandable in the context of that particular society, relating them to the coherent system of beliefs accepted by the people of that society. And on demand the concerned people can also produce or indicate a body of evidences in support of their beliefs. The evidence-relative character of their beliefs set limit to the subjectivity of the beliefs and somewhat insulate the same from the biases and prejudices of the individual human being. Close scrutiny of the evidences forming a coherent system and on which one is obliged to fall back in defence of his beliefs, theories and practices, etc. reveals that they themselves in turn, i. e. at a different level, are questionable and corrigible. Thus the history of science and technology of a society, questioned and corrected in the light of disturbing and testing evidences, changes, at times slowly and at times, rapidly over a period of time.

The questions and corrections affecting the course and career of science and technology need not be necessarily *internal.* They may be *external* as well. For the people of other society, belonging to other cultures, in spite of their initial handicap due to what may be called cultural distance, can and do understand the ideas and actions of our society. Contextualism does not mean cultural solipsism. Every society has its two complementary images—how it presents itself to itself and how it appears or projects itself to otherselves. The projected image may be an appearance in the sense that it may not have a point-to-point correspondence with the self-image of the society; but that does not mean that its personality is split. In other words, ontological unity and identity of a society is quite consistent with its dual or even multiple images. By implication this also suggests a limit to the thesis of constructivism which I accept partially and only partially.

The burden that new information generally imposes on a rational man is that the relation of the new information to the other organized informations with which the latter operates has to be defined, however provisional that definition may be. I draw a distinction between *knowledge* and *information* in this context. Knowledge is more demanding than information in the matters of seeking and accepting its place in one's world of beliefs and activities. As 1 have said, this is generally, and not, in all cases, true. For some people informations provided by a particular source or authority carries the weight of knowledge. "Who accepts whose authority?" illustrates a sort of question which cannot be answered in a universally acceptable form. The norms of acceptability cannot be defined in a context-invariant manner. The same point holds good in the case of the concept of rational man. There is no single or unique rational man referring to whom one can uniquely define the relations between information and knowledge. PS, MS and CS have their own models of rational man, relation between information and criteria of acceptability. Each one of these models, we must bear in mind, has logical room for several sub-models within it. The rational man, Magician, of PS may take a particular piece of information, '*i*', i.e. a message of God conveyed through a priest, quite consistent with what he accepts as a system of scientific knowledge, Magic, whereas the rational man, Scientist of CS may think either that '*i*' is inconsistent with his system of scientific knowledge, S, or that '*i*' and S are consistent provided *message of God, priest* and *conveyance* are interpreted in .a way which is consistent with S. In the latter case the acceptability of '*i*' is Contingent upon its interpretation satisfying the requirements of S. In principle '*i*' is, then, open to several interpretations answering the requirements of different systems of scientific knowledge, e.g. magic (M), Philosophy (P), Science (S) and Super-science (SS).

This line of my argument may lead one to believe that, in spite of my denial, I propose to defend a strong thesis of contextualism. The defender of this thesis, it is suggested, encounters lot of difficulties to answer such questions as "which one of the interpretations of a particular piece of information is to be accepted and which one of the available criteria actually availed of for the purpose?" Besides, such ontological questions as "what is there behind the information?" are also bound to raise difficulties. The major difficulty which bothers one is perhaps this. All these questions can be answered, broadly speaking, in two different ways: *internally,* i.e. from within a particular society and consciously bearing that society alone in mind, and *externally,* i.e. from within a particular society but bearing well in mind the views entertained on the matter by other societies and trying to understand them coherently. I prefer the term "understanding" to the term "rational reconstruction" because of its flexible logical connections. While I try to understand the ideas and activities of other persons belonging to other societies I cannot completely rise above the influences of the society to which I myself belong historically, linguistically, culturally and in various other ways. This relation of my belonging to my society can never be completely fathomed and clearly expressed. In this sense every answer to the above questions is bound to be internal. There is a limit to the sense in which a person can be exclusively preoccupied with the ideas and activities of his own society, completely ignoring those of other's. For every society is culturally more or less open-ended, i.e., without any *cultural-neutral protective belt* thrown around it, ensuring its autonomous existence and growth. Stated differently, *in real life situation cultural interaction is inescapable.* But there is significant difference between *entertained cultural interaction and imposed cultural interaction.* Under the conditions of imposed cultural interaction one has to react almost without freedom to external cultural stimuli. Consequently, the nature of the reactions tend to be clumsy, ill-formulated, vaguely or loosely connected with the rest of one's cultural acquisition. The external cultural stimuli to which one is obliged to interact are of diverse sorts, e.g. a flying saucer, a strange object, a talking tiger, an incredible story, an act of rising again to life of the dead, a strange episode, an infallible future-telling machine, an unbelievable "invention", a physician who can cure all diseases, a god who is all-knowing and all-powerful. From one's point of view each one of these sorts may appear extraordinary. Little reflection is necessary to convince ourselves that this is not necessarily the case. Even to what we regard ordinary external cultural stimuli our reactions are almost compulsive, although the consequences thereof on our system of ideas and activities may not be that conspicuous or serious. The entertained cultural reactions are in most cases anticipated selectively, orderly, well-formulated and more or less clearly connectable to the rest of our cultural acquisitions. If the full implications of what I call the inescapability of cultural interactions with or responses to the external stimuli are adequately realized, I think, the idea that I propose to defend a strong version of contextualism will be quietly abandoned.

The case and strength of contextualism as I understand will be clear if one carefully studies *how cultural interactions internally work.* Culture, like language, is an umbrella term. What it means are certain non-uniform, developmentally uneven and culturally interacting units. The relation of these sub-cultural units to the concerned overall culture is like that of the dialects of a particular language to the language itself. The borderlines of the sub-cultures are overlapping, interpenetrating and changing. Sub-cultures may be viewed in two different ways, vertically and horizontally. Vertically a culture, the Indian e. g., may be viewed as the totality of several sub-cultures, communities, of Bengal, Gujarat, Punjab and Tamilnadu, etc. Horizontally a culture may be understood as a totality of different strata of sub-cultures, of arts, morals, religion, technology, language, etc. Map of a language showing the relations and ranges of its dialects on it cannot be neatly drawn, at least not correctly, using only one colour or type of marking—vertical, horizontal or diagonal. One experiences comparable difficulties in drawing a map of a culture showing correctly the relations and ranges of its sub-cultures within it. What is even more difficult to show by drawing is the tie that binds the sub-cultures together. Political administration may be referred to as a well-recognized unifying tie. But 'history abounds with illustrations showing impotence of polity to define inter-subcultural relations. Culture has often been found to have defied and crossed the boundaries of polity. Equally difficult is to show the *depth dimensions of culture,* e.g. aesthetic, ethical and religious, encompassing all the sub-cultures. These difficulties are inherent in the very character of culture. Needless to say, the act of drawing a map of culture or language is itself a cultural act. Nature knows no mapping, no border or boundary on its geographical surface. Even "physical boundary" is a human artefact.

The interactions of the sub-cultures between themselves generate a *conflict-cooperation situation.* The unevenness of the depth dimensions also generate a similar situation. Industrial culture promotes individualism, petrifies inter-personal relations, and creates the conditions of alienation. The *dominant force* in CS is industrial culture. Agrarian culture lays emphasis on the community life, favours the joint (i.e. molecular) family system and face-to-face relations, and frowns upon the individual's right to question the community ethos. The dominant force in MS is agrarian culture. At times industrial culture is taken as a synonym of city culture and agrarian culture of village culture. Almost all cultures are composite in the sense that some of its sub-cultures are dominantly agrarian and some others dominantly industrial. The dominant forces of PS are generated by hunting, fishing and fruit-plucking. As I have said before, the same society is often found to be an unplanned locus of different cultures, different sub-cultures, each known by its dominant force or the *mode of production.* The dominant mode of production is not the way to understand the identity and function of a culture. At certain stages the cultures of a society are better understandable in terms of religious movement or warfare. *The role of ideas in* shaping the course and career of a culture must not be assimilated under *the role of the mode of production.* The interaction between these two roles is not one-sided. The *normal dominance of the mode of production* is occasionally interrupted by a novel and very influential idea or a set of ideas.

Ideas are both products and producers of culture. The course and contents of a culture, normally expressed in its ideas and ideals, are at times questioned, criticized and corrected by *newly found ideas.* When the culturally inherited ideas and ideals of a society fail to satisfy the growing *needs,* material and intellectual, of a sizeable section of its members, an objective situation is created necessitating the search for discovery and acceptance of new ideas and ideals. The said needs are generated by interaction between culture and culture, i. e. externally and, also, perhaps more so, by interaction between the sub-cultures, i. e. internally. The needs are expressed in the forms of problem and even crisis. In PS (T1) the external stimuli prove really disturbing, almost disruptive. But, in the absence of *culture-neutral protective belt* around it, PS (TI) has to learn to live with the "disturbances" of the external origin; gradually in the process partly it adjusts itself to the stimuli, partly it internalizes the latter, and partly it resists and rejects the same.[5] Having learnt how to respond to the external stimuli and 'having internalized the messages of the external cultures, MS (T2), and particularly CS (T3), are better equipped in terms of materials and intellectual apparatuses than PS (TI) to turn its critical and pointed attention to the problematic features of internal stimuli produced by the interactions between the structurally heterogeneous and functionally uneven sub-cultures. The veiled comparative evaluation of the cultural acquisitions of PS (TI), MS (T2) and CS (T3) as implied in my preceding sentence shows my own cultural grounding or point of view which, as I have already said before, may be concealed only upto a stage but cannot be logically disowned. However, I would like to add, this does not amount to giving up of contextualism as I understand it.

Openness of a particular culture of a society to external stimuli produced by Particular cultures of other societies sets a limit *to its autonomy* and to the extent of its autonomous growth. History and analysis would show that different sub-cultures, e. g. religion, polity and science, are not equally open to external stimuli or what may be called occasions of criticism. The sub-cultures less open to external criticism are generally found to be also less responsive to internal criticism. In the scale of stimulus-response or that of successful interaction, experimental sciences like physics and chemistry come up at the top, religion occupies a bottom position and polity figures somewhere in between. The sub-culture of a society, for example India, can hardly be indifferent to the newly established theories of other societies, like USA and USSR. Science as a spectrum of sub-cultures of different societies spread over the whole world is a community by itself. One might say that scientific culture of today in CS (T3) is really international, cuts across the political boundaries of different nations or societies, and the problems of translation are felt least in this area of men's cultural acquisition. Given the common rules of the game of science, i. e. its methodological uniformity, abstractness and "hegemonism", one might perhaps rightly assert that in the next few centuries to come, in FS (T4), the character of science would acquire more homogeneity and many other branches of knowledge, not yet known as strict science, would be brought under its expanding scope. Polity as a sub-culture of a society, like India is less responsive to the structures and functions of the polities of other societies, e.g. UK, USA, China and USSR. Not that interactions between these polities are not taking place; but compared to sciences their level is lower. Polity-wise human societies are somewhat conservative. In respect of religion these are even more conservative. While I am saying all these, one can easily find out my own point of view, i. e. the cultural background of my mind in CS (T3) which, partly accounts for comparative placements of science, polity and religion in the scale of interaction indicated above.

It is absolutely necessary to bear this point in mind. Otherwise one might be deluded to believe that the position of science, polity, or religion defined in terms of its openness or lack thereof has been alike in all societies—PS (TI), MS (T2) and CS (T3). The demarcation between science and religion in PS (TI) was not as sharp as we find it today in CS (T3). Not only in PS (T1) but also in MS (T2) polity has a close relation with religion: in CS (T3) this relation is becoming increasingly loose and formal. Another point which we should bear in mind is that at times following external aggression or annexation, the victor's polity is formally imposed on the total culture, including such sub-cultures as religion, polity and science, of the vanquished. I say "formally" because the imposition in the case of culture, especially in its non-material aspects, cannot be *effectively total*. The room for interaction, retreat or withdrawal is always there. The victor in one respect of culture, e.g. political organization or warfare, may through interaction prove vanquished in another, e.g. religion or science. The sub-culture which dominates a society in war generally cannot dominate it in peace. Those who wield arms—military leaders—dominate during warfare but yield their position to the masters of ideas during peace or act according to the latter's advice, i.e. allow them to play their role in effect. The basic features of conflict-cooperation situation are there even in the relation between culture of the victor and that of the vanquished. It is never a case of all-conflict-and-no-cooperation or of all-cooperation-and-no-conflict. The *grey area* of *cultural interactions* marked by varying degrees of assimilation, resistance, rejection, modification-and-acceptance is almost universal and very instructive.

It is interesting to note why a working scientist of a society at a particular point of time wants to resist the acceptance of a particular piece of information as true, even though he recognizes that it satisfies the logical and experimental tests accepted by himself. This is so because he realizes that its implications may immediately disturb and later on irreparably damage his "disciplinary matrix" or "set of paradigms".[6] Unless an information embodies the result of *crucial experiment—crucial* to the accepted set of paradigms—it cannot be taken as a threat to the set of paradigms as a whole. Informations obtained from that sort of experiment brings about a decisive or *revolutionary change* in the history of science. But revolutionary changes do not take place very often. Even the informations warranting a revolutionary change in the conceptual framework of science may be interpreted in a *conservative* manner in order to minimize their "damaging" or "destructive" effect on the concerned framework. To what extent crucial informations can change the *conceptual framework* of science depends, besides the nature of interpretation, *conservative* or *radical,* on the nature of the social background of science, i.e. whether it is PS (T1) or MS (T2) or CS (T3), and the logical rigours which define its systematic character. Certainly in PS (T1) and MS (T2), *logical system of science* did mean something different from what it does in CS (T3 or will do in FS (T4). Science as a sub-culture in PS (TI) or MS (T2) was much less autonomous and systematized than it is in CS (T3). Systematically incorporated in philosophy and religion, science and its logic shared the strength and weakness of the said two disciplines in PS (TI) and MS (T2). Let us not forget that the separateness of these disciplines is the result of the evolution of culture and increasing intellectual divison of labour.

REFERENCES

1. Wilhelm Windleband, *A History of Philosophy,* Vol. 1. Harper and Row, New York, 1958, pp. 23-24.
2. F. A. Von Hayek, *The Counter-Revolution of Science,* Free Press, Glencoe, Illinois, 1952, p. 30.
3. Michael Du Dummett, *Truth and Other Enigmas,* Duckworth, London, 1978, pp. 365-74.
4. Edmund Leach, "Models of Man" in William Robson (ed.), *Man and the Social Sciences,* George Allen and Unwin, London, 1972.
5. Imre Lakatos, "Methodology of Scientific Research Programmes" in Lakatos and Musgrave (eds.), *Criticism and Growth of Knowledge;* Cambridge University Press, 1972.
6. Thomas S. Kuhn, "Reflections on My Critics" in Imre Lakatos and Alan Musgrave (eds.) *Criticism and Growth of Knowledge,* Cambridge University Press, 1972, p. 271.

8

SCIENCE-PHILOSOPHY DIALECTIC

Do science and philosophy interact or not? The question has been answered both positively and negatively. Regarding the needs of man-in-society, some realist thinkers believe the nature and growth of science is clearly intelligible. There are others who thinks that both science and philosophy can well be understood as secular disciplines. A section of thinkers of the second group of ultra-platonic persuasion, has even claimed sovereignty for philosophy, trying to keep it free from "the disturbing day-to-day details" of science. This proclaimed freedom has an unwelcome implication: while the history of science is marked by progressive or regressive shifts, that of philosophy only by the repetition of the same old themes dressed up in new idioms.

One version of this pro-metaphysical concept of *philosophy* is found in Aristotle and Hegel, and another in Kant. Collingwood's attempt to clarify the relation between science and philosophy or metaphysics in terms of what he calls *absolute presupposition* is highly ingenious and represents an intermediate position between Kant's and Hegel's, drawing upon both. According to Aristotle and Hegel, the details of science are confusing and meaningless unless these are viewed under higher general *principles:* and the cognitive ascent ends up in God, which provides complete and the best possible intelligibility both to the details and the principles.[1]

Kant's philosophy is essentially a metaphysic of scientific experience. Assuming Newtonian mechanics and Euclidean geometry for granted, he, by regressive analysis, claims to have discovered certain presuppositions. To *justify* the necessary, unified and inter-subjective character of the scientific paradigm of his time, Kant refers to the ascending and convergent series of synthesis—in sense, in imagination, in understanding and finally, in the transcendental apperception of the self. The self is thus called upon to do the double duty (i) of *freely* uniting the elements of experience and (ii) *of receiving* the same (as spatio-temporally determined) from without, from not-self. Buchdahl observes :

> Kant certainly thought it a vital achievement to have exhibited putative links between the formulations of Newtonian mechanics and the transcendental principles of experience in general; moreover the former historically no doubt conditioned the general construction of the architectonic of the Critique.[2]

Without an all-designing God to support it from within and behind, the self cannot be expected to do its double duty and philosophically justify the paradigm science of the period.

Collingwood also wanted to justify and understand science in terms of some absolute pre-suppositions (APs), which themselves though metaphysical and non-propositional in character, enable us to determine the truth-values of scientific propositions.[3] To take away the absolutistic overtone of Aristotle and Hegel and the ring of finality associated with Kant's *Critique,* and perhaps primarily to explain the *growth* of science, he attributed a historical character to the APs, *relativizing* them to the scientific investigations of the concerned different *periods.*[4]

I put the question this way: Is it a *set of specific theories of science* (SSTS), quantum mechanics of Planck and Einstein, or what may be called *science as a sub-cultural whole (SSCW) of* the contemporary period, taking all types of science—natural, life, formal, social—within its scope?

The pre-suppositions of SSTS are different from and less general than those of SSCW. The pace and character of their change also differ. The experimental fortunes or test-consequences of quantum mechanics or relativistic physics have had considerable feedback and dialectical impact on the presuppositions of the concerned SSTS. But the same cannot be said of SSCW, for its pre-suppositions are very broad in scope, highly general and abstract in character. The relation of the latter with the former is "logically weak": these pre-suppositions cannot justify science but only help us to *understand it.* Strictly speaking, SSCW has very little test-consequences; and their feedback and dialectical impact on its abstract and general pre-suppositions is weak, slow and, in the short run, almost imperceptible.

But when a historian of culture gives us synchronous account of SSCW and its presuppositions thinly spread over and silently working through all other sub-cultures of the concerned society, one needs some middlemen, interpreters, to "show" us the "almost invisible" logical connections between the two areas. In these days of specialization, unfortunately, neither the philosopher nor the scientist ordinarily like this job of middleman. And it is only of late being taken up by scientifically trained historians (like Duhem, Needham, Wolf, Sarton and Singer) and philosophers (like Poincare, Russell, Popper and Kuhn). Given typologically their social and discourse-contexts, the presuppositions of various levels may be shown in the following way :

Primitive Society	*Medieval Society*	*Contemporary Society*
Time (1): PS (T1) "Magic" (M) as "Science" (S). Formal science discourse: FSD. Natural science discourse : NSD. Social science dis-course: SSD. Common sense discourse : CSD FSD/NSD/SSD/CSD High-level presuppo-sitions: HLP. Middle-level presuppositions: MLP. Low-level pre-suppositions: LLP	Time (2): MS (T2) "Philosophy" (P) as "Science" (S). Formal science discourse: FSD. Natural science discourse : NSD. Social science discourse: SSD. Common sense dis-course: CSD. FSD/ NSD/SSD/CSD. High-level pre-suppositions: HLP. Middle level presuppositions: M LP. Low-level presuppo-sitions: LLP.	Time (3) : CS. (T3) "Science" (S) as ". . ." Formal science discourse: FSD Natural science discourse: NSD. Social science discourse: SSD. Common sense discourse: CSD. F SD/ N SD/SSD/ CSD. High-level presup-positions: HLP Middle-level presuppositions: MLP. Low-level presup-positions: *LLP.*

Unless they are questioned about the meaning or truth of their thought or actions in CSD and, to a lesser extent, in SSD, people do not ordinarily refer to the concerned presuppositions. Compared with common sense and social science discourses, those of natural science or philosophy is clearly marked by problem-consciousness and anticipation of criticism. CSD is marked by three or four levels: (i) question and answer or description of facts, (ii) norms or forms of the concerned social context, and (iii) the principles which, if necessary, are cited to justify or to explain or to make level "ii" intelligible. Let me make the point clear by a simple example :

Level 0 : X drives his car at 50 m.p.h. along the Chowringhee Road (Caclutta).

Level 1 : Traffic police notes down the car no. and files a case against X before the appropriate authority.
Level 2 : X is punished.
Level 3 : Evidently X had no safety consideration for the human life.

One may like to bracket level-1 and level-2 together, Question may be raised (level-0) : "Why should X not drive at 50 m.p.h?" Answer may be : "Then X is liable to prosecution for the violation of traffic law...." (level-1), "Then X may be punished for..." (level-1) and (level-2). Answer may even be a straight forward (level-3) : "Safety consideration demands speed limit", assuming (level-1) and (level-2) for granted and, therefore, leaving the same unmentioned. Level-determination is discourse-relative. HLP of CSD (level-3) of the case, e.g. is open to may questions. Even being itself a justification of an answer to many other questions than the one mentioned at (level-0) above. HLP (level-1) of a SSD or moral science discourse (MSD) (Level-3), "Safety consideration demands speed limit," may be questioned by a social or moral historian on the ground that it is merely a *form* of prescription and *not* a prescription by itself. The genuineness of its historical character remains questionable unless those specific conditions are concretely spelled out which would satisfy the said form and be recognized as safety neasures in a society at a given time. Recognized safety measures expected to be (but not necessarily) followed by the car-driver, the traffic police, and the judging authority obviously undergo change over a period in response to technological, administrative and other reasons.

NSD is also marked by a comparable presuppositional level distinction :

Legel 0 : Reports of sense-data.
Legel 1 : Reports of observation about objects.
Legel 2 : Empirical general truths about objects.
Legel 3 : Exact and universal mathematical physics.

This may be made clear by an example:

Level 0 : Perception of series of [concentric] coloured ares [in a direction opposite to the sun].

Level 1 : Concentric coloured arcs with red atone edge and violet at the other and all other colours arranged in between them are observed when light from a distant source falls upon a collection of water drops, such as in rain, spray, or fog.

Level 2 : The white light of the sun refracted through raindrops shows the dispersed and separate colour corpuscle or the visible spectrum of the sun.

Level 3 : All physical phenomena, including the optical ones like rainbow, are (explainable by) the action of force representing attraction or repulsion, depending only upon distance and acting between unchangeable particles.

The rainbow has been rendered intelligible by the presuppositions of Newton's (1643-1727) *corpuscular theory of light* (MLP, Level-2) and the *mechanical view of the universe* as a whole (HLP, Level-3). It is both historically and conceptually interesting to recall that the same (rainbow) phenomenon has been rendered intelligible by a different presupposition (M LP, Level-2), viz. the rays of different wavelengths of different colours, though have the same velocity in the ether, differ in their velocities in raindrops and are separated. HLP, Level-3, of Huygens is also mechanical but not Newtonian. His reported understanding of rainbow (Level-1) presupposes the so-called Huygens' principle of secondary wave fronts, and which is claimed to be independent of his mechanical hypothesis (level-3).[5] If the secondary wave front principle is itself a level-2 principle and independent of the level-3 mechanical hypothesis, then certainly their relation begs some explanation. Otherwise the anticipated charge of adhocism becomes inescapable.

In the absence of clinching arguments or, preferably crucial experiment, two or more theories (together with their presuppositions) explaining the same phenomena can "peacefully coexist" as did Newton's corpuscular optics and Huygens' wave optics for nearly a century and half. And during this sort of period the choice of theory becomes, to quote Einstein, "more a matter of taste than of scientific conviction." It is to the credit of Newton's and Huygens' scientific conviction that they both agreed on a possible crucial experiment, which, if and when devised, would clinch the issue. If it could be experimentally shown that light, which, according to Newton, travels in straight lines and, passed through obstruction, casts sharp shadows, is capable of bending and, passing through a sufficiently small obstacle, casting no shadow, then that would have meant the refutation of the corpuscular theory and the vindication of the wave theory. This precisely anticipated falsifier of the corpuscular theory was brought to the notice of the scientific community by Young and Fresnel in the early nineteenth century.

The wave theory solved certain problems and, as it happens in the history of science, created certain others. Einstein and Infeld rightly observe :[6]

> In the attempt to understand the phenomena of nature from the mechanical point of view, throughout the whole development of science upto the twentieth century, it was necessary to introduce artificial substances like electric and magnetic fluids, light corpuscles, or ether. . . The principal physical ideas (of classical mechanics) have left for us unsolved problems. . difficulties and obstacles which discouraged the attempts to formulate a uniform and consistent view of all phenomena of the external world ... Modern physics has attacked all these problems and solved them. But in the struggle for these solutions new and deeper problems have been created. Our knowledge is now wider and more profound than that of the physicist of the nineteenth century, but so are our doubts and difficulties.

Drawing numerous examples from the history of physics, Einstein has shown us clearly how the solutions of some problems have given rise to "new and deeper problems," how widening and deepening of our knowledge result in wider and deeper "doubts and difficulties." According to Popper's scheme P_1TT-EE-P_2..., we start with some problem, P_1, move to a tentative theory, TT, which cannot be completely free from errors; then we try to detect and eliminate its errors, EE, by criticism and experiments; elimination of some errors gives rise to new problems, P_2 and does not mean the end of all problems.[7] Our main concern here is to see how philosophy comes into the historical picture of science: whether it is *presupposition* of science and anticipates the possible findings of the latter.

First, one is advised to bear in mind that PS (TI) and MS (T2) recognize no division between science and philosophy,[8] and that even in CS (T3) many pro-science philosophers, though draw a line of *demarcation* between the two, continue to recognize their *dialectical* continuity. In the past societies (what we call these days) science and philosophy are found to have coexisted in a state of uneasy peace under the canopy called *knowledge (Vijnana:* Sanskrit). The uneasiness was due to new discoveries proving inconsistent with the accepted "system" of knowledge, creating *problematic* situations marked by dialectical tension of cooperative-conflicting forces of creation and elimination. If systems of knowledge are sought to be constructed in a purely *a priori* way regardless of *problematic* experiences, available and possible, we are free to construct *opposite* theories. This is, in essence, the burden of Kant's argument in "Transcendental Dialectic" of the first *Critique.* This was Kant's way of refuting "transcendental metaphysics" and *not* philosophy of science. Hegel sought to rehabilitate "metaphysics" by his brand of dialectic, which welcomes opposition as promise of change and glorifies contradiction as harbinger of development, both of ideas and facts. A disgusted mathematician, David Hilbert, commented: "The thought that facts or events might mutually contradict each other appears to me as the very paradigm of thoughtlessness." Hilbert wrote well after Marx without apparently pursuing the latter's clarification that thought-process is not autonomous and reflects in a transformed way the dialectics of life-and natural processes. Marx finds no *self-realizing* dynamic in nature paving its way to the levels of life and thought and, therefore, speaks of *dialectical interaction* between the different levels of existence, of primary shaping forces emerging from the down below, the material process. He claims to have put the Hegelian dialectic upside down.[9] Merleau-Ponty's reading of Hegel-Marx controversy on dialectic seems to me eminently sensible :

> It is true, as Marx says, that history does not walk on its head [thought], but it is also true that it does not think with its feet [matter].[10]

All sorts of human knowledge, including philosophy and science, are certainly influenced by history and the social context, but the latter, in turn is also influenced by the former. Positively speaking dialectic means, as Popper says, the method of trial and error, the method of learning from mistakes (one's own and others'), of eliminating errors through criticism and test, and of getting closer to the truth.

Numerous presuppositions of knowledge are, broadly speaking, of three sorts—empirical and scientific, logical and mathematical, and metaphysical and philosophical. Philosophers like Quine do not recognize the proclaimed sortal distinction between the propositions supposedly constituting the presuppositional framework. Philosophy and science are continuous with one another, and so are science and common sense. This view clashes with Carnap's account of the difference between science and philosophy. While the scientist uses such words as "electron" and "positron", says Carnap, the philosopher only talks *about* such words. Spelling out his "strategy of semantic ascent" Quine rejects Carnap's metalinguistic characterization of philosophy and argues that we can both talk *about* the respective merits, e.g. of the wave theory and the corpuscular theory and also use the theories if we discuss whether corpuscles and wave exist, providing Quineans and Carnapians alike a common ground to communicate. Object-language and meta-language are logically inseparable and interanimated parts of one and the same biosocial framework. It is only within a socially workable system of language, ill or well-defined, that sense-data, objects, laws and theories of nature, and even categorical frameworks are identified and communicated. I reproduce three insightful quotations from Quine on the point :[11]

> "Science, though it seeks traits of reality independent of language, can neither get on without language nor aspire to linguistic neutrality."
> "Neurath has likened science to a boat while, if we are to rebuild it, we must rebuild plank by plank while staying

> afloat in it. The philosopher and scientist are in the same boat."
>
> "The philosopher's task differs (from the scientist's and the mathematician's only) in detail; but in no such drastic way as those suppose who imagine for the philosopher a vantage point outside the conceptual scheme that he takes in charge. There is no such cosmic exile."

Besides Popper, it is Quine who, among the contemporary English-speaking philosophers, has most successfully highlighted the developing (but, of course, "conservative") character of knowledge. Popper supports the strong insistence of Marx and Engels that "science should not be interpreted as a body of final and well-established knowledge...but rather as something developing, progressive."[12] Although he perfers a trial-and-error description of scientific development, marked by both tradition and revolution, to a dialectical one of the same, he himself admits that "this criticism is not of great importance".

If the Popper-Quine (continuity) approach to the history and philosophy of science is basically correct, then there is something wrong with Strawson's (unity) approach to the matter. (For the limited purpose of this paper I ignore the difference between Popper and Quine on many issues.) Strawson is against the *continuity* or historical approach to philosophy, by which he means analytic discovery of the basic conceptual (unity) framework of thought buried under the common sense concepts, and rejects Collingwood's and Körner's historical interpretation of Kant's presuppositions of science. According to him, "general scheme of thought" is changeable only at the peripheral ends while its core remains historically unchangeable. To quote him :

> " [C]ertainly concepts do change... certainly... metaphysics has been largely concerned with changes... but it would be a great blunder to think of metaphysics only in this historical style. For there is a massive central core of human thinking which has no history."[13]
>
> "[S] chemes of thought, employed by human beings reflect their nature. . . needs . . . situation. . . are not static. . .allow

> of that indefinable refinement, correction, and extension which accompany the advance of science . . [P]eople may. . . conceive of variations in... their own situation and needs and discuss intelligibly the ways in which their schemes might be adapted to such variations. But... conceivable variations are intelligible, only as variations within a certain fundamental general [unity] framework of ideas."[14]

Strawson's position stands or falls in between anti-historical Descartes and Kant, on the one hand, and pro-historical Marx, Popper and Quine, on the other. Strawson's "scheme of thought" is claimed to be both unified and anti-transcendental (even in the context of metaphysics). How can anti-transcendentalism and anti-historism go together? How a historical "massive central core" of human thinking can be logically deemed to be *unified* with the indefinitely corrigible and *historically* changeable areas? Strawson cannot satisfactorily explain the relation between the core areas and the peripheral areas of the *one and the same* general scheme of thought. Dualistic tension of his thought is unmistakable. Either he should assert, like Husserl,[15] for example, that the meaning of history is founded upon "the universal historical *a priori*" and "a universal teleology of reason", or call upon the "person" to do the double (and, I suspect, impossible) duty of *unifying* the changeable with the unchangeable without invoking the pro-Kantian concept of transcendental self. Any serious thinker who recognizes impact of science on society and thought, and, simultaneously, realizes the inadequacy of the positivist's attempt to explain both the known and the knowable exclusively in terms of the known would readily appreciate Strawson's dilemma.

But the way out, I submit, does not lie along Strawson's ambivalent lines of thinking. One must recognize that to explain even the ordinary objects of experience like colour and shape, not to speak of extraordinary objects like capacities of nature, planetary motion, and the rise and fall of an empire, presuppositions of different kinds (FSD/NSD/SSD/CSD) and levels (HLP/MLP/LLP) are inescapable. *Philosophy works not only as the presuppositions of science but also as its postsuppositions.* Prevalent worldviews of different kinds and levels are bound to influence the scientist and, in turn, his own logical and empirical findings may confirm or infirm those pre-critically accepted views. True, logical and empirical investigations may be undertaken with one of the two different ends in view—justificatory or critical: but it has to be admitted that the scientist *qua* man, in spite of his professional competence, cannot totally rise above the ideas and beliefs of his time. While analysing the metaphysical foundations of modern science, Burtt observes that one of the best ways of discovering the world-view of any age "is to note the recurrent problems of its philosophers [who] never succeed in getting quite outside the ideas of their time so as to look at them objectively—this would, indeed, be too much to expect."[16] To the medieval thinker man was the centre of the universe and nature subservient to his knowledge, purpose and destiny, and to the modern thinker, it is said, perhaps in an overgeneralizing vein, that nature exists and operates in its own independent way and man's relation to it is dependent on and determined by it. Because of the dominant philosophical and religious ideas of the time, not only the medieval European thinker but also his Indian counterpart took upon themselves the intellectual responsibility to see to it that the "scientific" findings of the time do conform to the officially recommended or traditionally accepted broad framework of thought and action. From Descartes onwards science in Europe is marked by the rise of mechanical, mathematical and measurable study of the natural world. Burtt observes:

> These founders of the philosophy of science. . . tended *more and more* to avoid (metaphysics), *so far as they could avoid it; so far as not, it became an instrument for their further mathematical conquest of the world.* (emphasis in original)

Records show that non-conformist views had been there always in the history of the philosophy of science, e.g. the views of Hobbes and More tend to limit "mathematical exploitation of nature". Comparable position in contemporary physics has been taken from the opposite end by Wheeler, who, following Clifford's monistic ontology and Einstein's unified field theory, affirms *"physics is geometry"*, despite the dominant influence of quantum mechanics marked by the uncertainty principle and field-particle duality.[17]

These examples may or may not be conclusive, but Burtt's indirectly admitted point stands, viz., in scientific search for truth, metaphysical conjectures, though at times sought to be avoided, are in fact unavoidable. To go beyond the gotten, the fact-fetish of the radical empiricist, the official tenets of the time, and to explore the structures of the yet unknown, metaphysical conjectures are indispensable. The point has been persuasively argued by Popper. He thinks that a particular problem-situation obliges a rational thinker to imagine one or alternative possible solutions, which to start with, may appear metaphysical or even mythical. "They may, in the course of discussion, become fruitful and important for science. (There are, examples of myths which have become most important for science, among them atomism and corpuscular theory of light"[18] According to Osiander, the heliocentric system of Copernicus was, to start with, nothing more than a "beautiful mathematical hypothesis" capable of better (than the Ptolemaic one) connecting the observable celestial phenomena. Experimentally not yet established Einstein's Unified Field Theory may easily be and in fact has been dubbed as metaphysical. But that does not necessarily take away the scientific significance of it or for that matter of any problem-oriented metaphysical theory. The points to be considered when entertaining a metaphysical theory are (i) whether it purports to solve a problem or a family of problems, and (ii) whether the proposed solution is criticizable and corrigible.

Influenced by and presupposing philosophy, science, in turn, dialectically influences and postsupposes philosophy. Attempts to draw a sharp line of demarcation between the two break down. Advances in science do change philosophers' world-view; and, as Harris puts it, "the appropriate form of philosophy for the second half of (this) century" is bound to be based on "what science tells us" and its implications. To quote him :

> The need... remains for the metaphysician's effort to see things together, as Plato recommended not to correct, outdo, or modify the pronouncements of science, but to reflect upon them, to develop their implications and mutual connections, examine their presuppositions, and to form as completed and systematic a conception of the world as the available evidence permits.[19]

Burtt has shown us the metaphysical foundations of modern science, while Harris and Buchdahl have demonstrated how "philosophical heritage is deeply grounded in, if not originating from, speculation about science," but, unfortunately, none of them have critically studied the dialectic between science and philosophy. This dialectic may be approximately represented by a series of *non-uniform* progressive epicycles. In a way Popper's scheme, P_1-TT-EE-P_2... gives a rough picture of the situation. Clash between knowledge and its background or presuppositions, on the one hand, and the relevant newly found facts, create a problem-situation necessitating a critical review of the theoretical picture and redrawal of it. Harris's observation that "we cannot step out of the charmed circle of knowledge... and (that it) must be self-reflective and ... self-validating" is somewhat misleading. Theory of scientific knowledge is not a "charmed circle," but, as I see it, a series of non-uniform progressive epicycles. The circle may be broken up both from within, by exhibiting logical inconsistency between the theoretical propositions or their implications, and from without, by discovering inconsistent observable facts. Sartre[20] speaks of critical *de-totalization* of the propositions and their dialectical *re-totalization*. The Sartrean totalization, unlike the Aristotelian or the Hegelian one, goes on without any totalizer or God. Neither philosophy nor science can live in peace with inconsistency *and* yet develop. Inconsistency presupposes a critical logic of relation between theoretical terms and propositions, on the one hand, and *observables,* on the other, and in the absence of which Hegelian dialectic, for example. easily gulps down and digests inconsistency, and the Duhem-Quine thesis can hold any statement true come what may, making drastic enough adjustment elsewhere in the system of science.[21] Popperian dialectic is *error-eliminative* and *truth-approximative* : but one can easily give it an unintended pro-Quinean twist and a bulgy and ugly *error-incorporative* look. For *truth-approximation* is an ambiguous concept. Marxian dialectic, in its critical role, is error-eliminative and, in its constructive role, combinatorial, trying to combine the seekingly irreconcilable extremes. Anti-dialectic prejudice tends to push a platonic metaphysician to an essentialist view of scientific laws which are said to be true, come what may in and through experience.

One or two examples may illustrate the point.

Plato's philosophical theory of Form itself, which laid the foundation of Euclidean geometry, owes its origin to the then problem situation in Greek science, especially Democritus' atomism, which followed the discovery of the irrationality of the square root of two, and purported to reconcile the motionless world-view of Parmenides with the fluxist worldview of Heraclitus. The Pythagorean theory of number, representable by dot-diagrams, contains the rudiments of a very primitive atomism. But both Pythagoras and Democritus, unaware of the proof of the irrationality of $-\sqrt{2}$, i. e., the incommensurability of the diagonal of a square with its side, thought that every measurement is reducible to pure numbers. Once given, the implication of this proof shook the very foundations of their theories, for both of them thought that every measurement is reducible to pure number. This proof dashed down the cherished hope of deriving cosmology, or even geometry, from the arithmetic of natural numbers. To save the Greek science from this crisis Plato developed his theory of Forms, an autonomous geometrical method, freeing mathematics from the "arithmetical" assumption of commensurability or rationality, and anticipating the Elements of Euclid.[22]

A radical empiricist may unwittingly characterize Plato's theory as purely speculative and criticize it as fanciful. But Charles Singer, the distinguished historian of science, has very rightly observed that "fancies of this type have been repeatedly of value in the history of science."[23] Without the theories of motion developed by Buridan and Oresme, considerably improving upon the Aristotelian impetus theory, the development of cannon, a break-through technological innovation, would have been impossible.

In the history of Indian science and philosophy too one comes across the problems of reconciling the motionless world-view of the *Vedantins* and the fluxist world-view of the *Madhyamikas*, the observables or multiplicity and the universals or unity and the recognition of the irrationality of $\sqrt{2}$.[24] The *Nyaya-Vaisesika* theory of atom is a bold speculative attempt to explain how partless and eternal atoms are brought together in the composition of objects by two forms of external motion, desert (*adrsta*) and shock (*samksobha*) entailing impact (*samskara*) and velocity (*vega*) produced by some conscious agency aided by God.[25] But it appears that the dialectic between the speculative Indian theories of motion and the mathematical-mechanical study of observable natural objects could not develop to the extent it did in Europe. The ancient Indian geometers, Apastamba and Baudhayana, are often credited to have first found out the proof of the irrationality of $\sqrt{2}$.[26] An important question yet to be answered is why in spite of remarkable achievements in the fields of mathematics and astronomy (between B. C. 500-950 A. D. and as evident in the works of Arya Bhata(s), Bhaskara(s), Brahmagupta, etc.) we did not have in India a major mathematical-physical breakthrough comparable to the one in Europe between 1300-1700, from Buridan to Descartes and Newton? Maybe it is due to the inadequate recognition of the significance of the mathematical-mechanical approach and also of the necessity of subjecting metaphysical speculations to its rigorous demands.

Those who have carefully studied the striking similarity between the philosophic-scientific ideas of the ancient Indian, Greek and Chinese thinkers can hardly miss their dialectical development. Clash between (newly observed) facts and (existing) theories, partial assimilation of the former in the latter, and the resulting modification and enrichment, revolutionary (Popperian) or otherwise (Kuhnian), of the latter in the light of the former are the basic traits of this dialectic.[27] Quest for knowledge, philosophical and scientific, is *interaction* between the two and not onesided action of the one upon the other. Interactions between and within the cultures also largely contribute or, at times, arrest this dialectic. As there is no sharp distinction between theoretical terms and observational terms, there is no sharp line of demarcation between the cultural and natural factors underlying the philosophy-science dialectic. While cultural factors may prove even *coercive,* natural ones may fail to be more than merely *suggestive* in their impact on theory-change. This seeming anomaly is due to the often unrecognized anthropological fact that the knowing man, primarily a product of one culture and yet open to several other cultures or/and subcultures, is engaged in producing another culture according to his needs and understanding. Marx says, humanism and naturalism are dialectically intertwined." This dialectic knows no threshold or ceiling and its operational modalities are multiple. This view may also be given an unintended Duhem-Quine twist. However, generally speaking, the problems and solutions of man-culture and nature-culture relations exhibit comparable patterns because of the basic unity (or the laws) of nature in its operation. Sarton observes :

> Many Greek ideas in science and philosophy are duplicated in India (and China) . . . The duplications help to prove the essential identity of the human mind (shaped by culture and nature). Given definite problems that admit of only a few solutions (because of the law-governed unity of nature), it is not astonishing that wise men of Greece, India, China, etc. hit independently upon the same solution (s).[29]

The philosophy-science dialectic of Europe has been thoroughly traced and differently interpreted by many distinguished scholars. Joseph Needham and his associates in their monumental work, *Science and Civilization in China,* have related this dialectic to the broader and more illuminating framework of the mode of production and technological development. India is yet to have her badly needed Needham, Sarton and Thorndike. Only some such thinkers as B.N. Seal and D.D. Kosambi[30] have left behind some seminal ideas on the close relation between philosophy and science in India. Unfortunately for us, most of the competent scholars who studied Indian philosophical texts have taken an exegetical, historical and a somewhat dreary approach without particularly caring to discover their scientific implications and cultural presuppositions. To understand the philosophy-science dialectic of India what is called for is an inter-disciplinary approach, and, needless to say, classical scholarship should be a very important component of it. The collaboration between the professional philosophers and the working scientists would be the central point of this very promising research-programme.

REFERENCES

1. *Philosophy of Nature,* tr. A. V. Miller, Foreword J.N. Findlay, OUP, 1970, p. XXIV; see also, pp. 6-13. Findlay says, "Hegels's view of Nature is. . .

carrying to the limit of an immanent, Aristotelian teleology, in which Nature is to be understood as throughout working towards an end which will ultimately carry it beyond itself."

2. Gerd Buchdahl, *Metaphysics and the Philosophy of Science,* Basil Blackwell, 1969, p. 678; see also pp. V. 512-30.
3. R.G. Collingwood, *Essay oil Metaphysics,* OUP, 19.10.
4. Stephen Toulmin, "Conceptual Change and the Problem of Relativity" in Michael Krausz (ed.) *Critical Essays on the Philosophy of R.G. Collingwood,* OUP, 1972, p. 218. Toulmin has rightly pointed out that Collingwood: 'left it uncertain whether he intended his term. [APs] to apply to the most fundamental theoretical principles within a science, or to the disciplinary principles constitutive of it."
5. John Herival, *The Background of Newtonian Principia,* OUP, 1966, see also, M. Minnaert, *The Nature of Light and Colour in the Open Air,* Eng. tr., new edn., Dower, 1959.
6. Albert Einstein and Leopold Infeld, *The Evolurion of Physics,* Cambridge University Press, 1971, pp. 120-21; see also p. 112.
7. Karl R. Popper, *Objective Knowledge: An Evolutionary Approach,* Clarendon Press, Oxford, 1972, p. 119.
8. E. A. Burtt, *The Metaphysical Foundation of the Modern Science,* Routledge & Kegan Paul, London, 1964. See also, E.E. Harris, *The Foundations of Metaphysics in Science,* George Allen & Unwin, London, 1965; and M. Capek, *The Philosophical Impact of Contemporary Physics,* Van Nostrand, Princeton, 1961.
9. C.D. Karnig (ed) *Marxism Communism and Western Society,* Vol. 2 Herder & Herder, New York, 1972, pp. 394-420.
10. M. Merleau-Ponty, *Phenomenology of Perception,* tr. Colin Smith, Routledge & Kegan Paul, London, 1970, p. XIX.
11. W.V.O. Quine, *World and Object,* MIT Press Camb,1975, pp. 222, 223-24, 275; see also J.J.C. Smart, "Quine's Philosophy of Science" in Davidson and Hintikka (eds.) *Words and Objections,* D. Reidel,1975, p. 4; and Rudolf Catnap, *Philosophical Foundations of Physics,* Martin Gardner (ed.), Basic Books, New York, 1966.
12. Karl R. Popper, *Conjectures, and Refutations: The Growth of Scientific Knowledge,* Routledge & Kegan Paul, London, 1969, p. 334.
13. P. F. Strawson, *Individuals: An Essay in Descriptive Metaphysics,* Methuen & Co., London, 1959, p. 10.
14. P. F. Strawson, *The Bounds of Sense: An Essay on Kant's Critique of Pure Reason,* Methuen & Co.; London, 1966, p. 44; see also pp. 118-21.
15. Edmund Husserl, *The Crisis of European Sciences and Transcendental Phenomenology,* tr. & intr. David Carr. Northwestern Univ. Press, 1970, p. 370, see Kant, *On History,* ed. & tr., Lewis White Beck, 1963, pp. 11-26.
16. E . A . Burtt, *op. cit.,* p 1; see also, pp. 2 4, 10-11, 302-05.
17. A Grunbaum, "Space, time and geometry" and C.W. Misner, "Some topics for philosophical inquiry concerning the theories of mathematical geometrodynamics and of physical geometrodynamics" in *Boston Studies* XX, PSA 1972, (eds.) K . F. Schaffner and R . S. Cohen, D. Reidel, 1974.
18. Karl R. Popper, *Conjectures and Refutations, p.* 257; see also *Logic of Scientific*

Discovery, Hutchinson, 1959, sec 85, p. 278; J.W:N. Watkins, "Metaphysics and the Advancement of Science," *Tire British Journal for the Philosophy of Science,* Vol. 26, No. 2., 1975.

19. Errol E. Harris, *op. cit.,* pp. 29; see also, pp. 22, 489.
20. Jean-Paul Sartre, *Critique* of *Dialectical Reason,* London, 1976, pp. 59, 502-04.
21. Sandra G. Harding (ed.), *Can Theories be Refuted: Essays on the Duhem-Quine Thesis,* D. Reidel, Amsterdam, 1 976.
22. Karl R. Popper, *Conjectures and Refutations,* pp. 75-93. See also Charles Singer, *A Short History of Scientific Ideas* to 1900, Oxford, pp. 25-27; and Ernest Hutten, *The Origins of Science,* Allen & Unwin, London, 1962, pp. 68-71, 126-41.
23. Charles Singer, *op. cit.,* p. 24.
24. Satya Prakash, *Founders of Sciences in India,* New Delhi, 1965. See also, D. M. Bose *et al.* (eds.), *A Concise History of Science in India,* New Delhi, 1971.
25. Umesh Mishra, *Conception of Matter,* Allahabad, 1936. Needham tells us that "there are reasons for us to believe that the roots of the theory of *Paramanu* (atoms) go much further (than the Vaisesika) back in the history of the Indian thought".
26. Bibhutibhushan Datta, *The Science of the Sulba,* 1932. See also Bibhutibhushan Datta and Avadesh Narayan Singh, *History of Hindu Mathematics,* 1962.
27. Imre Lakatos and Alan Musgrave (eds.), *Criticism and Growth of Knowledge,* Cambridge University Press, 1970; see also, Nicholas Maxwell 'The rationality of Scientific Discovery", *Philosophy of Science,* Vol. 41, No. 2, 1914.
28. See e.g., German *Ideology; Holy Family, and Economic and Philosophical Manuscripts* quoted from T.B. Bottomore and M. Rubel (eds.) *Karl Marx,* Watts & Co., London, 1956, pp. 54-56, 245-46.
29. George Sarton, *A History of Science,* Harvard Univ. Press, 1966, p. 255. See also Aifred Weber, *History of Philosophy,* tr. Frank Thilly and R.B. Perry, New York, 1925, p. 22.
30. See B.N. Seal's *The Positive Sciences of the Ancient Hindus;* and *Science and Human Progress: Kosami Commemoration Volume* (eds.) D.S. Kothari and D.R Gadgil, Popular Prakashan, Bombay 1974.

9

Towards Understanding an Alien Culture

The relation between science and philosophy in the past was not how we view it today. The meaning of *science* and *philosophy* has also undergone significant change over the known types of societies—primitive society at time 1, in brief PS (TI), medieval society at time 2, in brief MS (T2), and contemporary society at time 3, in brief CS (T3). This process of changing relation is bound to spill over in future society at time 4, in brief FS (T4). The paradigm of "science" and that of the "wiseman" also change, differ from society to society. In PS (T1) the wiseman was known as magician (M); in MS (T2) natural philosopher or *philosopher* (P); and in CS (T3) *scientist* (S). The paradigm of wisdom in PS (T1) was theology-cum-magic; that in MS (T2) natural philosophy or philosophy; and that in CS (T3) is science. In FS (T4) it will be, let us say, super-science (SS). We should bear in mind that the names of society and paradigms we propose to use are typological and not descriptive.

In this paper I would like to show (1) that our *understanding* of other societies, their paradigms of wisdom and wisemen is objectively grounded in our own paradigms which are either consciously chosen or pre-reflectively accepted; and (2) that the logic and psychology of choice/acceptance are shaped by a very complex process of social interaction marked by both conflict and cooperation between and within cultures and sub-cultures. In this connection to clarify my view (3) I would very briefly refer to two important case studies often mentioned in the histories of science: the concepts of (a) irrational number, and (b) atom in ancient India and Greece.

It would be interesting to try to understand the reasons leading to the gradual independence of science. Philosophical reflection and analysis of history of science reveals that speculative, mythical and experimental elements got mixed up in it. This mix-up may be viewed either as logical and systematic or prelogical and presystematic. Why one views the composite elements of speculation, myth and experiment in one particular way rather than another largely depends upon one's own point of view. The concepts of logic, logical laws and system are not identical in different societies. Also widely different are the techniques and formulations of experimental findings. What was once recognized as a *logical system of knowledge* in PS (T1) is perhaps regarded today, in CS (T3), as an *indefinite whole of ill-assorted informations, beliefs and fancies* brought together by untestable speculative links. Even the basic laws of logic "basic" to "us"—do not convey the same sense to the widely different cultural groups. The things which present themselves distinctively different to the "naked" eyes may appear basically same to the "tutored" mind. The questions such as "whether eyes see in effect nakedly?" and "whether mind can act or be applied in a totally untutored way?", are not peculiar to the philosopher of CS (T3). Needless to say, the proposed answers to the questions, closely analysed, reveal their different cultural parenthood.

The said cultural difference seems to operate at two different levels, primary and secondary, or to put it differently, *descriptive* and *interpretative.* The levels may be more but not less. By the *primary levels of intelligibility* I mean how a wiseman, Magician, of PS (TI), understands and describes his own society, i.e. from within. By the *secondary level of intelligibility* I mean how a wiseman, Scientist, of, say, CS (T3), understands and describes PS (T1) or MS (T2). How a wiseman, Philosopher, of MS (T2) understands and describes PS (T1) may be a matter of understanding and description for a wiseman, Scientist, of CS (T3). Thus more than one level of understanding or several levels of understandings may be simultaneously object(s) or matter(s) of one's understanding. A history of ancient science and philosophy written by a contemporary historian is almost bound to be different from one written by a medieval historian. The difference is due not only to the *informative contents* but also to the *conceptual frameworks.* The conceptual framework with which a historian works has an important say in the determination of the contents and also in the definition of the scope of history in question. It may be pointed out that at every level the historian is obliged to work with one or other conceptual framework for the purpose of organizing informations and also perhaps which are believed to be informations. And because of this basic feature of thought even at the primary level, the informations, correct as well as fancied, which a historian claims that he is "objectively describing" are found to be mixed up with interpretation. Interpretation, however. does not mean distortion of facts beyond recognition. The question of "distortion" is generally raised by those naive realists who believe that "pure facts" can be seen by "naked eyes" and "pure informations" understood by "untutored or clear mind." The contribution of the conceptual framework and that of *the given* can be separated at least for the limited *analytic purpose.* The separation, however, can never be clear-cut because of the continuous dialectical relation between the two.

As there cannot be a circumference without a point, there cannot be a view without a point of view. This may sound trivial. But it is not. Once we closely look into the dialectical relation between what for the analytic purpose we split up into two and designate one as "conceptual framework" and the other, in the absence of a better expression, as "given," we may discern different levels of "operation" or "working" of the said relation. The level-difference may be observed also in the relation between the eyes and what the eyes see. The ocular structure of *normal* human beings is same, yet their observation reports vary, differ and at times even conflict. For the time being let us forget the difficulties involved in the attempt to define "normal human being". Frankly speaking I do not know whether the ability to see of all normal human beings is same, irrespective of their different ecological factors. I strongly suspect it is not. The power of vision of the people used to bright electric or tube light and that of the people used to wick lamp or no lamp at all after the sun set are perhaps not the same. Habit and experience are not likely to leave one's vision totally unaffected. Secondly, even if it is assumed that the ocular structure is same, the question remains whether that alone determines the qualities of its *functioning.* Does the "naked eye" see at all? Or is it the "clothed eye" which sees? Don't memory, attention and expectation enter into one's act of seeing? Someone who is born blind and then gets back his vision as the result of some operation cannot see, order and organize the objects as we do in the first few days. With the same ocular structure, one might say, "he does not see what we see." Bearing in mind the factors involved in the act of seeing it seems advisable to use some such expressions as "somebody sees" rather than "eyes, naked or clothed, see." Eyes are used by us in various ways, actively and attentively, passively and indifferently, to express or communicate some idea or emotion to impress someone or other, and so on. In a way the whole personality of a man *orients* his act of seeing. Thirdly, the dialectical relation between the seer and the seen has its level-difference and which seems to me very instructive. Sometimes, the seer "dominates" the "show" and the physical identity of the seen is "distorted" almost beyond recognition. I say "almost" because if the "distortion" is total then the term *recognition* makes little sense. Besides the term *distortion* itself begs elucidation. Sometimes disortion is deliberate and intended for some or other creative purpose, e. g. an artistic one. The seer's deep-rooted desire, intense dislike or hatred, obsession or complex, of which the seer himself is not ordinarily conscious, also colours the identity of the seen as borne out by others' observation reports about the latter. An element of distortion is perhaps inherent in the situation, in the very act of seeing. The seen-in-itself cannot be perceived in its entirety by the seer, however passive might be his posture. The limits are set not only by the psycho-physiological background and abilities of the seer but also partly by the geometrico-physical properties of the seen. The seen can hold out (so to say) its identity and unity even in the face of very active and dominant posture of the seer. The latter cannot wish the former out of existence without deluding himself or, stated differently, except in bad faith. In its dialectical relation to the seer, I feel like saying, the seen has a life of its own which may be changed but cannot be totally destroyed by the former.

Finally, the seer-seen dialectic has a very important bearing on the relation between the cultural *objects* and *how* they are understood. If the wiseman of CS (T3), whether he is an anthropologist or a historian of science, wants to understand the science and culture of PS (TI) directly and in terms of his own conceptual framework, i.e., totally or partially disregarding how the wiseman of PS (TI) understands the *objects* of science and culture from within his own society, he is bound to be in serious difficulties. His descriptive account of what he has heard and seen appears incoherent, bizarre or even meaningless and absurd. If the second level intelligibility of S of CS (T3) does not take due cognizance of the first level intelligibility (of M of PS (T1), S(scientific)—account of PS (T1) would be somewhat like the description of his surroundings given by the man who was born blind and has recently gained vision following operation. The description of the seen ("objects") does not necessarily make sense. For description itself is concept-mediated and not pure. Without being *physiologically blind* a person may well be *culturally blind.* A man who has right cultural vision of CS (T3) may be unable to "see" the cultural objects of PS (T1) or of MS (T2) i. e., may fail to grasp the intended meanings and relations—intended by the concerned people—of what he sees. Even then the main reasons which explain inter-cultural communication, meaningful interaction and mutually intelligible translation of their languages area (a) the *continuity* of different societies and (b) the *identity* of the world (of objects) we all, in a sense, commonly share.

Physical objects like "table", "chair", "tree", and so on are obviously different from cultural objects like marriage market, solving a problem and so on. This, however, is not to deny that the latter has some perceptible features which are not essential to the understanding of their meanings. From the other end one may point out the *physical* objects are not bare given either. The line of distinction drawn between them can neither be permanent nor straightforward. For instance, one may always argue that, in spite of its *physical* features a chair or a table is an object of *culture.* The same can be said of a tree as well if it happens to be a hybrid variety cultured by some known or unknown people. If providential teleology is allowed to have its say in defining an all- encompassing whole of objects, even mountains and rivers may be declared culture-objects *at bottom.* Though extreme, it is an easily conceivable position; and history abounds with its examples of different variety.

"What is the correct way of understanding an alien society,

say, PS.(T1)"? may be represented by another question "what is the correct way of translating the language of an alien society, say PS (T1), into that of our language, CS (T3)?" As we are aware, language is used for different purposes—e.g. to make a statement to express a belief to deny what is, ascribed to or associated with somebody, to request, to evaluate, and so on. The diverse purpose which a language serves and the ways how it serves, are closely inter-related and indefinite. A language which lies otherwise idle or almost dead and is good only for the limited purposes of recording or preserving some past states of affairs, ideas or practices can by acts of mouth or body be brought back to life and put to use of different sorts. Because of the interwoven character of language and society and also because of the innumerable ways in which each society uses its language, the task of the culture-translation often proves extremely difficult and delicate, except in the limited cases of description of one's bodily states and of perceptually identifiable medium-sized physical objects.

When we speak of the interwoven character of language and society, we imply, among other things, that the nature of culture objects is very much influenced by the language using which these objects are designated. Expressions of culture-concepts are under standably even less rigid. "Marriage" as a *name of* a culture-object is certainiy more rigid than its *description* "a form of union between a male and a female." But contextual analysis of their meanings will show that none of these designators is absolutely rigid. All over the world "marriage" obviously does not mean an identical group of observables or stimuli: overlapping is unavoidable. Nor the relations responsible for the grouping together of the said observables or stimuli are same. The "forrns of union between..." which could be truly predicated of "marriage" are on all ethnological accounts, numerous and widely different. "Chair" is certainly a more rigid designator than "marriage", although it occupies, in our context, a borderline position between physical objects and culture-objects. For an inter-culture translator or interpreter, therefore, "chair" is likely to pose much less problems than "marriage". To give a description of "chair" certifiable by stimuli is easier than to give a description of "marriage" certifiable by similarity. The scope of stimulus-certificate in the former case is comparatively limited. But the difference continues to be one of degree. However, to the culture-translator the most serious difficulties are posed by the objects of belief, worship, praise and blame. We can well imagine of a PS (T1) in which gods, angels, demons, etc. are all objects of true belief and worship, and the concerned people have their *own accepted ways of justifying* those beliefs and acts of worship. We can also imagine of a PS (T1) or a MS (T2) in which gods, angels and demons are all objects of belief but, while gods and angels are loved and worshipped, demons are feared and propitiated in different exotic ways, and they too have their conventional ways of justifying their beliefs and acts accordingly.

When from the area of knowledge we move to that of evaluation, the elements of indeterminacy and therewith the difficulties for the culture-translator increase. One possible way of circumventing the difficulties has been to draw and highlight a questionable distinction between the meaning of a term and the ways its reference is given. The description of marriage by which the reference of "marriage" is given in a PS (TI) may not apply to the intended reference of the term in (other possible worlds), say, MS (T2). By postulating a meaning of "marriage" which is equally neutral to all languages-and-societies we may think of a logically possible way out of the culture-translator's difficulties. But that proposed solution has its own difficulties. In abstract theoretical contexts postulation of intensional objects is permissible or perhaps even necessary primarily for "constructionist" purposes, but in the cognitive sociological context it seems unnecessary, if not misleading. Whichever society is taken to be the "basic" locus of the meaning of "marriage" its language and other associations are bound to enter into the very ways of its being given (as an object of understanding). To deny this entails an indefensible form of essentialism. It is difficult to imagine the identity of a culture-object totally disregarding the ways of its determination. If the object is conceived as a purely ontic entity and in a strictly non-epistemic context, one finds no harm in it. But one wonders whether there is any interesting point in taking culture-objects within the scope of a modal operator to indicate its non-epistemic status of rigidity or flexibility. For to the culture-translator or the historian of science perhaps the more interesting question would be to study how it is determined, i.e. who speak(s) of it; where, when, how and in which language it is spoken of. In other words, his main concern is the social character of meaning, i.e. *how* a particular sense is attached to an expression by the group of people who use the language in which the said expression figures. And this important point, relatively neglected by Frege and Russell, has been fortunately brought to the focus of our attention later by Wittgenstein and Quine.

Whether science and philosophy can be satisfactorily *demarcated* remains a controversial question. Carnap and other positivists at one stage argued that philosophy is a meta-theoretical activity and that it does not add to our knowledge of facts obtained otherwise from science. The main aim of philosophy is to make use of logical language for the purpose of the construction of a worldview accepted by the scientific community. Proper use of logic in that case can enable the philosopher to keep science pure or to purge it off; the pernicious effects of the pseudo-problems resulting from metaphysical speculations. It is well-known that Carnap extensively revised later on this view on the relation between philosophy and science. But, given this view, the right aim of the historian of science would be to relate the story how the scientists in course of time, step by step, relying on observations, devising suitable experiments, and using more and more "powerful" formal languages, have succeeded in freeing the "essence" of science from its speculative trappings.

One very basic question is left unanswered by this approach and, that is, why it is so that the "essence" of science itself has been persistently exhibiting a historical character. Is it not a fact that what are called these days philosophical *speculations* have often favourably *influenced* the course of *science* by enabling the scientists to frame testable hypothesis for the purpose of organizing the available experiences and anticipating the possible ones? Have not often scientific discoveries sparked off different speculative hypotheses for the purpose of rationally organizing the available as well as the possible experiences? Disregard of these basic questions and the underlying *dialectic* between the speculative elements and the empirical ones in man's cognitive enterprise explain to a great extent the unfortunate but the instructive fact that during the hey days of logical positivism while we all heard so much of admiration for science no significant history of science was produced by any leading spokesman of the movement. Exclusive preoccupation with the formal language of science, forgetting its parentage, i e, ordinary language and its social context, made many a positivist somewhat blind both to the roots and fruits of science.

To undo the effect of this fancied cleavage between science and philosophy who did most in the recent years are Popper and his followers, on the one hand, and Quine and his followers, on the other. For Popper's philosophy is basically scientific, consists in framing bold and problem-oriented conjectures which through criticism and test mature into scientific or other empirical theories. This process of maturation is uneven and unending.

Popper is prepared to accept only to a limited extent Wittgenstein's stricture against philosophy that it has no problem of its own. But that does not mean that philosophical problems are not genuine. May be their roots are elsewhere in science, politics, religion, or some practical needs. It is misleading to distinguish "studies" or "disciplines" by the subject matter which they investigate. The same subject matter may be studied and used to solve different problems. Popper speaks of three sorts of reasons defining the scope of different disciplines: (a) historical, (b) administrative, and (c) theoretical—theories designed to solve problems. For a proper understanding of the relation between science and philosophy, and that between society or history, on the one hand, and philosophy or science, on the other, one has to look mainly into (a) historical reasons, and (b) theoretical reasons. According to Popper, the philosopher does not merely "talk about philosophy," he tries to solve the problems both theoretical and practical, linguistic and factual. Since philosophy does not generate the problems generally known as philosophical but which result from the interaction between philosophy arid other disciplines, their *degeneration* or *development* is contingent upon the character of the interaction between the philosopher and, for example, the sociologist, the theologist, the scientist and the mathematician. And this is evident from the history of philosophy and science, taken both separately and jointly.

One of the main reasons used by the earlier positivists for the exclusion of philosophy from the domain of science is that the so-called philosophical statements are neither factual (synthetic aposteriori) nor logical (analytic apriori). That the analytic/synthetic issue cannot be easily decided *a la* the positivist has been persuasively argued by many contemporary philosophers like Heinemann,[2] Quine,[3] Watkins[4] and Putnam.[5] One and the same statement or set of statements may be interpreted both as factual and analytic. Popper refers to Newton's theory which, though generally recognized as factual, has been interpreted by Poincare and Eddington as implicit definitions. Unless a language is sufficiently formalized and its rules of interpretation or bridging principles are clearly formulated and illustrated, we cannot logically and satisfactorily decide which statements are synthetic and which ones are not. Popper and Watkins speak of some speculative statements, which are nonanalytic, criticizable (but *not* testable), and establish their factual *bona fides* by influencing science.

To illustrate the point, Popper refers to the ancient Greek thought—how the crisis in Pythagoreanism and early Greek atomism led Plato to lay the foundation of Euclidean geometry.[6] The point however was noted earlier, among others, by Singer and Sarton. Plato's philosophical or speculative theory of forms owes its origin to the then problem-situation in Greek science, especially Democritus' atomism, which followed the discovery of the irrationality of the square root of two. Pythagoras thought that all things of the world, including even the qualitative ones, are, in essence, numbers or ratio of numbers. The Pythagoreans, Aristotle writes in his *Metaphysics,* thought that mathematical principles are "the bases of all things" and that almost all things are numerically expressible. But, then, they also developed a conception of irrational quantities, not expressible by ordinary numbers. Singer has rightly pointed out: "with the imperfect mathematical notation of the time... great algebraical advance was impossible, and irrational numbers could not be algebraically, represented.' Greek mathematics was thus forced to preserve its geometrical bias."[7] However, this theory of number was successfully applied to simple geometrical figures such as squares, rectangular and isosceles triangles, to certain simple solids such as pyramids, and also to, abstract ideas such as Justice, Beauty, Harmony and Knowledge.

Though the Pythagorean theory of numbers, which can be

represented by dot-diagrams, contains the rudiments of a "very primitive atomism", Popper's conjecture[8] is that the atomic theory of Democritus was mainly influenced by the *theoretical* controversy on the problem of change between Heraclitus, on the one hand, and Parmenides and Zeno, on the other. Heraclitus' theory is simple: nothing is, everything flows; identity is illusion, change reality. Equally simple and attractive is Parmenides' theory : the world is one and full; it knows no void, has no parts, and, therefore, is motionless. Democritus rejects Parmenides' conclusion and having done that questions the premises as well. His view is the world consists of parts, is not full, i.e., the void is there, and, therefore, motion is possible. Motion is external to atoms and not within them. For Democritus' atoms are full, i.e. know no void within, and indivisible—somewhat like miniature replica of Parmenides" big world. Movement in the world is due to different possible arrangements of atoms. Democritus' theory of atoms has been, credited with the power of explaining such emipirically known properties as degrees of hardness and resilience, rarefaction and condensation, compressibility, coherence, disintegration and many others. It also anticipated the calculus of integration. "But perhaps", Popper says, "the most fascinating elements in Democritus' theory is his doctrine of the quantization of space and time." The proof of the irrationality of the square root of two, $\sqrt{2}$, was unknown to both Pythagoras and Democritus and its implication shook the very foundations of their theories, for both of them thought that every measurement is reducible to pure numbers. This proof "destroyed the hope of deriving cosmology, or even geometry, from the arithmetic of natural numbers." To save the Greek science of that period from this crisis, Plato developed his theory of Forms, an autonomous geometrical method, freeing mathematics from the "arithmetical" assumption of commensurability or rationality, and anticipating the elements of Euclid.

Of course, Plato's theory of Forms was speculative or, as a radical positivist might like to characterize it, fanciful. If a theory has to establish its "scientific" credentials in terms of its "being based on sense-experience", both the Pythagorean theory of Numbers and the Platonic one of Forms are bound to be expelled from the arena of science and criticized as fanciful. Singer has very rightly observed that "fancies of this type have been repeatedly of value in the history of science."[9] The radical positivist fails to correctly formulate the relation between the role of theoretical speculation and that of experiment and observation. "(T)he formation of general ideas on theoretical grounds has (often) preceded and not followed practical (experiment and) observation.'[10] The mystical Pythagorean view that the sphere is the perfect figure led most of the astronomers of the later ages to think that the earth as well as the planets are spheres. Tycho Brahe's (1546-1601) hypothesis of the Universe as according to the ideal form of the circle, is one of the last great reminders of the Pythagorean spirit. Kepler's (1571-1630) idea of Universe was also essentially Platonic and Pythagorean. Right from the beginning he was persuaded that the order of the Universe and its parts is in accord with some abstract ideal of the beautiful and the harmonious; and that it must be expressible in numerical and geometrical form. However, it was left to Descartes' genius to bring about the long-awaited union between the Pythagorean theory of number and the Platonic theory of form.(1627) His analytic Geometry was the point of convergence of Hindu algebra and Arab geometry and its application proved to be the greatest single step of the seventeenth century in the progress of the exact sciences. And this is evident from the scientific works of Pascal, Galileo, Torricelli, Huygens and several other eminent scientists of the time.

Interaction between science, i.e., observational-experimental quest for knowledge, and philosophy, i.e., speculative quest for the same, is not peculiar to Greek or European thought. The same theme with local variations can be illustrated by a rational reconstruction of Hindu thought of the ancient and the middle ages. George Sarton's view on this point is noteworthy: "Many Greek ideas in science and philosophy are duplicated in India. It is very interesting to compare these duplications... The duplications help to prove the essential identity of the human mind. Given definite problems that admit of only a few solutions, it is not astonishing that wisemen of Greece, India, China, etc. hit independently upon the sane solution."[11]

Necessity, practical as well as theoretical, is rightly said to be the mother of invention. Necessities are of different sorts, viz., cooking food, irrigating and cultivating land, measuring space (plots of land) and time (hours, months, years and so on). Over the years and in the light of organized experience necessities change and so do the forms of meeting those necessities. This is true both for the growing individual and the changing society. Our actions are in the nature of response to external stimuli or internal urge: stimuli or urge may invite effort to understand or attempt to solve problems or both. Annual floods in the Nile valley washing away the boundary lines between the plots of land and renewing the disputes between the owners of those plots proved to be a pressing practical problem to the geometricians (literally "measurers of land") of the ancient Egypt. The *Elements* of Euclid were in the main a theoretical response of far-reaching consequences to the practical needs of the time.

The origin of *Sulba-Vijnana* (literally "science of measurement") in India is connected with the construction of the altars of the Vedic sacrifices. Baudhayana and Apstamba, the two most outstanstanding geometers of the ancient India, are said to have flourished well before 500 B.C. Sulba meant not only "measuring" but also[10] the unit of measure" or "an instrument of measurement" which was a rope or chord *(rajju)* at that time. The ancient geometers of India spoke of three kinds of measure— linear, surfacial and voluminal—and five types of specialists in mathematics—expert in the Sulba or geometer, inquirer *(paripre-chaka)* into the *sulba*, and uniform rope-stretcher (*sama-sutra-nirancaka*). Democritus (470-500 B.C.) is also said to have used the term *harpedonaptae,* the Greek equivalent of rope-stretcher. Satya Prakash surmises that the geometrical ideas of India influenced Greek sciences.[12] The ancient Indian geometers such as Baudhayan and Apastamba did recognize the irrationality of the square root of two, $\sqrt{2}$. They had their proof of it as well.[13] Some scholars like Schroeder and Burk think that the credit for the first discovery of irrationals goes to the ancient Aryans of India. However, other equally competent scholars like Zenthen, Cantor and Vogt have expressed their doubt over the matter.

The *Nyaya-Vaisesika* theory of atom embodies another bold sepeculative attempt to solve the problems of unity and multiplicity, of identity and change. And the problems have to be studied against the background of controversy between the *Madhyamika Buddhist*, emphasizing the principle of *dynamics* or flux, and the *Vedantin*, emphasizing the principle of status or identity. Against the Nihilist school of the Buddhist, which holds that void (*sunya*) is the only real entity, Kanada, the propounder of the atomic theory, says that atoms *(paramanu)* are partless and eternal "reals". Motion (*karma*) is external to atoms and all the non-eternal objects are ultimately reducible to various atoms through motion. Atoms are brought together by two forms of motion for the formation of composite wholes: (a) desert *(adrsta)* of the human beings or creatures who are to make use of their body or object concerned, and (b) shock *(samksobha)* entailing impact *(samskara)* and velocity *(vega)*. Both desert and shock are externally produced on the atoms by some or other conscious agency helped by God. All objects consisting of atoms have space, *akasa*, as their locus or substance, an eternal continuum, and the latter, imperceptible in itself, is known inferentially through its quality, sound. To account for the relations of priority and posteriority, of simultaneity and succession, etc., and to measure the units known as moment, minute, day, month, year, etc. the atomist feels logically obliged to posit time, *kala*, another all-pervasive and eternal substance. In itself imperceptible, time is known through its determinations. Time which relates the objects of the universe with the movements of the sun, is instrumental cause of motion and, therefore, also of the production, existence and destruction of every product of atoms.[14]

It is very difficult to state with definiteness whether Indian atomists influenced their Greek counterparts or the latter influenced the former. Sarton's guess that "the Phoenicians, who were very clever dragomans and middlemen, may have transmitted some Hindu (atomic) theory" to the Greeks, may or may not be right. But his anti-diffusionist and pro-convergent conclusion on the matter appears very rational and accords well with a wider range of independently accepted historical facts: "...the Greeks were quite capable of reaching that solution (of the problems of unity and multiplicity, of identity and change, in terms of speculative atomic theories) by themselves, and so were the Hindus."[15]

In the face of same or similar problems, thinking human minds are repeatedly found to have discovered independently same or similar solutions in the past. The question, "why the thinking human beings face same or similar problems?", seems to me very interesting and a part of a larger issue which has to be studied in cooperation with the cultural anthropologist, the sociologist of knowledge, the linguist and the ontologist. Not that science cannot be understood as a secular subject, i.e. strictly in terms of its "technical" problems and solutions as viewed by the professionals *from within* their own community. But my concern here is with the history of science in its sociological perspective, or, to put the same thing somewhat differently, the *sociology* of science in its historical perspective. Whether it is taken as a *corpus of theories* or a *form of activities,* science is undeniably social in several complementary senses. First, basically it is a product of social cooperation and interaction and not of some individual's insight and excellence. Occasionally "accidents" do contribute to the career of science. Secondly, the set of symbols or the language used for *doing* science or for the purpose of construction of its *theories is* also social. The rules of interpretation of scientific language/activities are a part of social life. Thirdly, both the problems and the proposed solutions of science are social or public, i.e. open to common understanding and test. Problems result from the perception of inadequacy of an accepted conceptual framework to deal with some or other objectively recognized objects of experience or informations ascertainly grounded in nature and culture. Fourthly, the very development of science shows its objective and social character. The individual scientists who discover the problems of science and try to solve them are all mortals like ourselves, but what they leave behind, their works outlive them and may be understood without making any *specific* biological reference to them. Finally, *rationality (from within)* of science and its *intelligibility (from without)* provide yet another proof of the socially and historically continuous character of science. The considerations and proofs that I am offering to show clearly the social moorings and implications of science are interrelated, overlapping and convergent.

Before I conclude, perhaps one point, already mentioned, needs to be stressed. Once the nature/culture dualism, clearly untenable in the light of modern findings of such sciences as biophysics, physiological psychology and biolinguistics, is given up and, in the wake of that, it is duly recognized that rudiments of cultural problems are there in the natural level as well, we would find it easy to understand why historically distanced and geographically scattered groups of people came across similar problems and solved or tried to solve the same in more or less similar ways. From this one might hastily infer what Levi-Strauss describes as "cultural universals". Whether we speak of "natural kinds" or "cultural universals", one impression we must not give and that is of their *fixed essence* against the tide of time. Nature is diversely *individuated* by time. Culture is diverse because of the incurable *individuality* of the historically evolving men and groups. The said two sorts of diversity, are admittedly identical at *bottom* but too much of emphasis on "bottom" often leads one to metaphysical and sociological *essentialism* making one avoidably blind to the diverse and refined factors of social life.

REFERENCES

1. Michael Dummett, *Truth and Other Enigmas*, Duckworth, London, 1978, p 424.
2. F.H.D. Heinemann, *Proceedings of the Xth International Congress of Philosophy*, Reidel Amsterdam, 1949.
3. W.V.O. Quine, "Two dogmas of empiricism," *Philosophical Review, 60*, 1951.
4. J.W.N. Watkins, "Between analytic and synthetic," *Philosophy*, Vol. 32, 1957.
5. Hilary Putnam, "The analytic and the synthetic," in *Mind, Language and Reality: Philosophical Papers*, Vol. 2, Cambridge University Press, 1975.
6. Karl R. Popper, *Conjectures and Refutations*, Routledge & Kegan Poul, 1962, pp. 75-93. Here I follow Popper's account and interpretation of the problems and suggested solutions. Works of S. Sambursky and Mashall Clagett are noteworthy in this connection. The results of historical researches of Charles Singer (*A Short History of Scientific Ideas;* 500 to 1900, 1959) and of George Sarton (*A History of Science*, Harvard Univ. Press 1966) are substantially in accord with Popper's interpretation of the problem-situation at the time.
7. Charles Singer, *op. cit.*, p. 26.
8. Karl R. Popper, *op. cit.*, pp. 82-83.
9. Charles Singer, *op. cit.*, p. 24.
10. Charles Singer, *op. cit.*, p. 27.
11. George Sarton, *op. cit.*, p. 295. See also, Alfred Weber, *History of Philosophy*, tr. Frank Thilly and. R.B. Perry, Charles Scribner, New York, 1925, p. 22.
12. Satya Prakash, *Founders of Science in Ancient India*, The Research Institute of Ancient Scientific Studies, New Delhi, 1965, p. 605.
13. Bibhutibhushan Datta, *The Science of the Sulba*, 1932. See also Bibhutibhushan Datta and Avadhesh Narayan Singh, *History of Hindu Mathematics*, 1962.

14. Umesh Mishra, *Conception of Matter,* Allahabad, 1966.
15. George Sarton, *op. cit.,* p. 255; see also pp. 17-18. "Each settlement had its men of genius, its dullards, and its great majority of average people..."

Part III

Man-Society Interaction

Man-Society Interaction

Once we keep our focus steady on the evolutionary nature of man and its implications, the error of overemphasis on theory, relegating practice to a pliant secondary position, or the opposite error, may be substantially avoided. It is well known that certain views of Plato, Descartes, and Kant, irrespective of what they themselves intended, are being used by their modern followers to defend the "autonomy" of theory, value, and society. The basic presupposition seems to be that theory, rightly formed, is expressive or reflective of facts so unmistakably that it needs no modification or correction through practice, test or questioning. Similarly, one could claim that values, rightly grasped, and society, rightly constituted, admit of no change. If it does, it is degenerative. This anti-evolutionary view credits man with some native and *immutable competence* to know facts infallibly, to grasp faultlessly the values which do not change, and to form perfect society that would be stable, if not permanent. But the records of history and the results of researches in cognitive psychology and biology are at variance with this optimist view and seem to suggest that the said human competence itself is movable.

One might ask, against this anti-evolutionary view, "Is it really possible for man to have that seemingly superhuman competence, to know facts infallibly, to grasp changeless values, and to form ideal permanent society?" Another question often asked in this connection is, "In face of the fact that many views of such competent thinkers as Plato, Descartes and Kant have been proved to be wrong, would it be reasonable to hold that the human mind can get into the innermost laws or secrets of nature and culture? If the human mind is indeed so competent, what is the harm in exposing it to the testing details of practice?" By exposing our theory and ideal society to *critical* practice, it is suggested, let us *try* and *test* them. Let us see the outcome of the trials and tests. Why should one be afraid of facing reality? Why should we shy away from it?

The response of the defender of the "autonomous theory" is somewhat like this. Theory is constructed in order to explain facts, and if that requirement is satisfied, and if the theory in question is found to have strong explanatory powers, we must accept it and that is the end of it. The question of further testing it by exposing it to practice is declared to be superfluous. But is it so? A theory is not tested only by the "known" facts which it explains, or explains better, but also, perhaps more so, by the facts not yet clearly "known" and purported to be explained by it. Explanatory powers of theory, to our mind, are open-ended and therefore at no stage of its development should one proclaim that its worth has been *conclusively* established. The "worth" of a theory does not consist only in its explanatory power but also in its truth. Rightly understood, *explanatory power* and *truth* are intimately related. But, as we know, systematic attempts are being made, interestingly enough, both by rationalists and positivists to play down the role of truth and to emphasize the *coherence* of the facts explained. Their presuppositions are admittedly different. The positivist does it because, according to him, *coherence* is the *test* as well as the *nature* of truth. If anticipated facts, anticipated by theory, come off and are found to be coherent, the worth of the theory is proved. The question of its being true or false need not be raised separately. According to the positivist, theory is an *instrument* of action and anticipation. Either it pays, ill or well, or it does not, but the question of its truth or falsity appears *meaningless* to him. Habermas, among others, has criticized this view at length (in his book, *Theory and Practice*). The reason for the rationalists' accent on coherence is different. The coherence of the facts of experience has a deeper *meaning* for him. The *surface* coherence of empirical facts, he seems to *see,* is sustained by a *deep* coherence of reality itself. This privileged *rational* vision of reality is, in a sense, self-certifying and, therefore, given the coherence of facts of experience, he does not require any additional proof of its truth. In either case coherence turns out to be the terminal point of theory construction and the search for truth through practice is declared unnecessary and, therefore, not undertaken.

In essays 10, 11 and 12 of Part III, the main point I have tried to make is very simple: whatever man makes, inherits, or needs—

theory, values, or social institution—is questioned, enriched, or impoverished, and, in the process, changed. Merely by virtue of its power to give a coherent picture of facts of a particular domain of experience, a theory cannot escape criticism and question for a long time. Belief in a value, political or religious ideology, for example, can certainly bring and hold together some people for some time, but history tells us repeatedly that it does not work for ever, at least not in the same way. Similarly, it may be pointed out, supported by many examples, that the unifying or cementing role of a social institution, is bound to be questioned and changed over time. If this phenomenological or plain description of theory, value, and social institution highlighting their evolutionary character, is accepted, I do not know why coherentists, at any rate a large number of them, should show, as they do, a marked resistance against almost every *practical* plan of man to *transcend,* deliberate effort to go beyond, the given experience. It is not clearly understood that perpetual pupilage in the school of experience is not by itself very educative. "Natural" experience, or accumulation of experience without plan or programme, neither tests nor enriches a theory. Values are not "natural" gifts. Unrelated to human needs, biological and sociological, the nature of values is not determined in abstraction. Social institutions are also meant to answer certain human needs and, in that sense, embody certain values or, as at some critical periods of history, disvalues. The evolutionary character of theory, value, institution, etc. does not mean superfluity but, rightly understood, necessity—perceptive necessity—of the human intervention in the process. The acculturation of nature is not entirely natural, and therefore should not be left completely to the forces of nature : it is primarily a matter of human enterprise. Transformation of both nature and culture is brought about by human ideas and actions.

Pro-evolutionary attitude is not necessarily pro-revolutionary or even pre-revolutionary. But that is the misconception or apprehension one finds in some liberal political and economic theorists. They think, taking their cues of thought from Hume, Adam Smith, Bentham and Mill, that political and economic processes should be allowed to be shaped by the spontaneous forces of nature, including those of "free" human nature, without let or hindrance from the side of the state authority. The government which governs least and the economy which is left to the "natural" forces of supply and demand and not interfered with by the state planning authorities or some such "artificial" power are said to be best and most rational. Daniel Bell, for example, is opposed to the idea of framing ideology for, according to him, it involves going beyond sociology, the social situation, which can well tell us what is to be done and what is not to be done. Similar views have been defended by Friedman and Oakeshott. We do not need any abstract and artificial ideology for practical guidance. Experience is enough and the safest guide. Hayek, in a similar vein, defends Adam Smith and Mill and pleads for a very limited scope of government activities. Bell's *End of Ideology* and Hayek's *Constitution of Liberty* harp the same themes: no (or minimal) transcendence of experience and no (or minimal) interference with the natural activities and ideas of men. It has been claimed that Popper's position is different because of the emphasis it lays on the *criticizable* character of all traditions, scientific and humanistic.

In essay 10, "Theory and Practice," I have indicated the reasons why theory is and should be exposed to practice. Without this exposure theory is not tested, corrected and precisified. It is not a one-sided need. Practice tends to be aimless or, at best, *ad hoc* without theory. Theory-provides a perspective of, provisional guidelines for, action. Theory-practice dialectic is both a sociological phenomenon; it is there, and as an ideological need, it *should* be there. Man is a systematically ambivalent creature: he lives on what is available, and simultaneously craves for what is *not* available. The relations between theory and practice have been analyzed from an individual as well as collective standpoints.

In the next essay, "Satyagraha and Social Change", I have tried to develop the general thesis that man is the main dynamic of social change. It is with reference to the ideas of Gandhi that I have argued the main point. Underlying every ideological view on social change there is a theory, articulate or inarticulate, of human nature and its various competences and their limits. In his view on *satyagraha* as means of social change, Gandhi has laid more emphasis on individual dynamics than on group dynamics. Groups tend to be stereotypes: group activities, in course of time, tend to degenerate into lifeless imitation, mechanical repetition, and cease to be innovative and creative. This tendency or degeneration, Gandhi is convinced, can be substantially arrested, provided the potential power, the spiritual power, that is there in every individual, is activated and put to use. This is what Gandhi calls the moral force of social change. Though he has spoken of the necessity of mass or collective *satyagraha*, Gandhi always highlights the role of *individual* responsibility. That the moral force or the "soul-force" of *satyagraha* has significantly contributed to the social change in India between 1930 to 1950 can hardly be denied.

The very nature of social change partly manifests and partly presupposes two things : one, certain human needs and achievements necessitate change; and two, certain institutions representing certain *other* human needs and achievements arrest the same. Even the society which appears structurally and outwardly static does always have within it a dialectical tension. This is basically due to the presence and function of the *human* factor. At the secondary level this factor has its institutional spillover effects. And these effects do not work homogeneously. Because while one set of human needs—poverty, hunger, slavery, unemployment, rising expectations, etc. gives rise to or strengthens pro-change socio-political institutions, another set of human needs—expansion "of proprietory rights, vested interests in luxury and comforts, craze for security, fear of freedom entailing growing responsibility, etc.—promotes anti-change institutions. In brief, the human needs represent conflicting interests. In essay 12, "Sociology and Ideology of Rural India and the Indian Peasantry", I have enumerated and analysed certain sociological factors—tradition, religion, *karma,* rebirth, Hinduization or Islamization, etc. which, to my mind, are mainly responsible for the rigidity of the rural economic structure of India. I have also enumerated and analysed, briefly, certain factors—education, increasing industrialization in the country as a whole, extensive network of roadways and communications, new agricultural inputs and mode of production, new land laws—to show that a pro-change awareness is slowly dawning on the mind of the peasantry.

10

THEORY AND PRACTICE

I
CONCEPTUAL PRELIMINARIES

The word "theory" has not been always kindly taken and rightly understood. At times even an element of criticism or sarcasm is implicit there against the theoretician. For many people theory is a strange and difficult (unnecessarily made so) subject. The concept of theory is a suspect only to those who are not quite clear about its meaning and role. The etymological origin of the word conveying the concept is not only respectable but also close to commonsense and there is nothing strange about it.

Theoria in Greek means view or vision. The exact meaning of the concept is not unrelated to the sense of the original word. There are two different, but not quite unrelated, ways of looking at or understanding an object or a thing. One, we may look at it primarily as a *whole* relatively ignoring for the time being its constituent parts. In other words, we may understand an object as a *unitary fact* to start with, and relatively ignoring its constituent factors which could be taken up later on for detailed analysis. Two, our primary interest in understanding an object may be its constituent *parts* or *factors;* for the time being we might as well forget or ignore their *gestalt* or holistic pattern. Theory provides a unitary vision of the whole, but by no means does it imply that parts of the whole are unimportant—unimportant even from the theoretical point of view.

The suspect character of theory assumes credibility primarily because the theories which are constructionist in nature deriving their elements from discontinuous space-time stuff and depending for their validity on the application to the fields which are spatio-temporally discontinuous or discrete. The subject matters of theory may or may not be in one spatially continuous time-frame. To a great extent this determines the relative abstractness or concreteness of theory. It also raises or removes difficulties in the identification of the test-conditions necessary for the validation of theory.

Every theory has its two practical aspects. First, the very origin of theory may be easily, in some cases a bit circuitously, traced to some practical difficulties. I am using the word "practice" obviously in a wide sense. Not only *originally* but also *terminally,* practice is relevant to theory for lending or establishing the dignity and strength of the latter.

The word "practice" is used in two different senses, positive or value-neutral and normative or value-loaded. Pro-naturalist thinkers take it as an equivalent word for test or verification. Because of its general and at times even a little vague character, naturally theory has to be exposed to tests, repeated tests and rigorous tests. So far as the question of natural sciences is concerned it is quite possible to define and obtain the conditions, repeatable conditions, necessary for rigorous tests. Consequently, the difficulty raised by the subjective factors and value elements could be minimized in the case of theories of natural sciences. This is also partly true, but only partly in respect of life-sciences. The inadequacy of causal analysis and explanation and the necessity of teleological concepts have been frequently mentioned in this intermediary area of sciences even by those who are otherwise favourably disposed towards the pro-naturalistic methodology.

The question of value has been most persistently put across naturally in the field of social sciences. Concerned as they are with human motivation and action, it has been argued, the theories of social sciences can hardly afford to ignore the normative aspect or the value component of practice. Some thinkers have gone to the extremity of suggesting that social sciences, including linguistics and economics, are essentially praxiological. The sophisticated exercises for predictable measurement of utility and uniform structuralization of the rules of syntax and semantics are said to be condemned either to fail or be embarrassingly inexact.

II
INEXACT CONCEPTS OF THEORY AND PRACTICE

Concepts are basic components of theory. Logically and appropriately related concepts constitute statements. Theory is constituted by a set of logically related statements. Logical relations may be deductive, inductive or probabilistic. Needless to say, all theories are not logically pure or neat. Most of the theories consist of "loosely" related "network" of statements. The underlying logical (not necessarily rigorously formal) character of these loose relations cannot be easily gathered from the statements of theory expressed in ordinary or natural language. Barring the well-defined concepts of formalized systems, most of the concepts we use are more or less inexact. Their inexactitude is basically linguistic. Concepts are embedded in language and not isolates. Since in most cases concepts are understood by us in their language-entrenched form and not in their purity or as ideal types, it is an epistemic obligation to take language seriously in order to ascertain to what extent their inexactitude is due to their language-context.

We are placed in a rather curious situation. Without language concepts in their fullness of meaning cannot be grasped. But because of their language-relative character we cannot get to their exact meaning. Three responses are possible to the situation. One (a) we may stick to the view that conceptual inexactitude and also the resulting theoretical inexactitude are inherent in natural languages, and (b) argue that with the help of formal logic we can take care of this problem and get to well-defined concepts. But this approach rests on a wrong notion about the relation between the formal language in which concepts are sought to be ideally defined and the concerned natural language which is invariably required to provide the necessary sub-structure of the formal language. Two, concepts are abstract entities and their entitative existence are not language-bound although they may *appear* to assume slightly different meanings in the context of different *uses* of language(s). Besides these two, one could think of another response and to which I am personally inclined and that is somewhat like this. Admittedly without language, concepts cannot be used and understood; that does not mean that concepts have no *relative* independence of their own. To defend the thesis of relative independence of concepts one need not postulate the abstract entitive existence of concepts. For, among other things, that is patently inconsistent with the very idea of the formation and transformation of concepts. And once we reject that basic idea we are invariably landed in pure Platonism, in an overpopulated world where the relations obtained within the population, i.e. in between the concepts, are fixed and frozen for ever.

Concepts may be inexact, even independent of their language-relative character. Their inexactitude may be traced either to their speculative origin or because of no or little exposure of the theories (in which they figure) to practice or both. Strictly speaking, concepts are not put to test or practice in an isolated manner. Their fate depends on the tenability or otherwise of their parent theory. It is true that concepts in isolation are used for identification of objects. There the question we encounter is whether the identification in question is precise or not? At that stage ordinarily the question of judgement does not arise. If, of course, it is insisted upon that every act of identification involves an implicit act of judgement, then that is a different proposition.

Theory is indeed parental in character. It would be futile to search for the origin of theory along the traditional lines, empiricism and rationalism. Theory owes its "origin" neither to pure experience nor to pure reason but to less exact theories. Experience, reason and the objective world are all mixed up in the realm of theories. Analysis reveals certain distinguishable elements in theory (or in the world of theories), via (a) language, (b) concept, and (c) object (or the world of objects). In between (a) and (b) and also between (b) and (c) further analysis reveals other elements. For example, one might say that language enters into the very constitution of object and this happens "because" of experience. This view has been controverted and it is being said instead that concepts have their extralinguistic existence. But that raises a further question regarding the relation, if any, between the so-called extra-linguistic objects and an all-comprehensive system of signs and symbols together with their "natural" rules of interpretation, whether we call it a universal language or a "universal mathesis".

Concepts in their separate existence and significance cannot be vehicle or expressive of clear thinking or articulate thought. When concepts themselves are represented as fully expressive of thought, the underlying rules of their relation with other concept(s) and the concerned rules of interpretation, though remain undisclosed, are implicitly understood. Secondly, the concept-object relation is also not free from ambiguities. "Simple" concepts which can rightly claim the competence to represent "simple" objects do not raise much of a problem. But, as we all know, "simple" objects are not really simple; analysis brings out their complex or composite character; and their constituent elements are often derived from different places and times and in different ways, through experience, imagination and, maybe, intuition. The very admission of the existence of *composite* concepts amounts to a stricture against the Platonic notion of concept. By implication we take it that concepts are formed and the elements out of which they are formed are not necessarily found together in the world of things. Therefore, there is always a little uncertainty about the question whether concept does or does not represent thing or, at any rate, simple object. The analysis may be carried on a little further towards the world of things, even beyond that of objects. First, the old question about the relativity of simplicity and complexity of object may be reopened. Simplicity of an object is not a onesided affair. It also depends on how it is viewed and reviewed. how the concept in question (again embedded in a theory) and its language are used to identify, a particular *thing* or group of things in the world. In fact, there is an element of human contribution, contribution of the knowing mind and shared by a community of minds, in the constitution and formation of object itself, which is ordinarily, and perhaps rightly so, regarded as relatively independent. So, an adequate theory must take into account besides language, concept and object, the attending social situation and what may be called the world of things.

In a prereflective level of thought all these elements of an adequate theory are not duly taken into consideration. Consequently, we encounter a host of inexact concepts. Myths of different societies provide a classic example of the *use of inexact concept* and the *formation of imprecise theory.* Examples may be drawn also from the areas of child psychology. This is *not* to suggest that the logic of mythology and child psychology are of the same category. Nor does it suggest that the disciplines containing inexact concepts are not amenable to logical treatment. Inexact and imprecise theories also have their logical structure(s). For obvious reasons, these structures are relatively complex and marked by various intricate relations. While we talk of inexact concepts and imprecise theories primarily we have empirical discourses, as distinguished from strict logical ones, in mind. In the world of myths we find, for example, the concepts of *creation,* how chaos become cosmos, *transformation,* how immortals become mortals and the converse, and *causation,* how the curse or boon of gods and godly creatures can initiate and execute a course of events which cannot be confirmed by our laws of science. In brief, mythical concepts and theories, though purported to be empirical, often do not tally with our standard and testable human experience. From this one might hastily conclude that the inexact concepts of myths are all fancy and fantasy and products of unregulated imagination and therefore need not be taken seriously for the purpose of logical and semantic analyses. Less uncharitable views on the matter have been expressed by philosophers and anthropologists like Levi-Strauss and Malinowski. Myths have their identifiable reference which may not be ostensive or directly locatable. Mythical world is said to have its own reality and that reality, though not co-extensive and co-ostensive with the world of scientific experience, is continuous with and resembles partially or even substantially the world we live in. Inexact concepts are resemblance-classes. The objects to which they apply may not clearly and completely illustrate them. Consequently, the theories containing inexact concepts or resembling predicates cannot be completely verified. An element of apriorism or whatever name is given to it has to be tolerated. But this is no peculiar fault of mythical concepts and theories. Even in the empirical sciences when one is to deal with objects which are continua, processes, dynamically complex, or historically fluctuating in character, one cannot logically claim to have had appropriate and exact concepts which would fully reflect them.

The world of myth full of inexact concepts raises some fundamental questions for those who are firmly committed to the ideal of the unity of theory and practice. If theories incorporate inexact concepts, their own scope become indefinite and character imprecise, and consequently their practical test can never be rationally satisfactory, unless test and *rational satisfaction* are loosely defined.

In this context the concept of magic is very relevant. The role of magic is conceptually necessary to bridge the practical gap between the inexact concepts of mythology, on the one hand, and the necessary conditions for their actual test on the other. The powers of magic and the magician are often called for and invoked in the mythical world to narrow down the gap between "theory" and "practice" and at times even to make the impossible possible.

III
APPROXIMATIONS TO EXACT CONCEPTS AND PRECISE THEORIES

Obviously the modern scientific mind is not prepared to accept the position that by magical power it is possible to remove the gap between inexact mythical concepts and the necessary but non-existent test conditions. When I say this I am not attaching any special importance to "the modern scientific mind", except in the limited historical sense that in relation to the mind of the past or the primitive man equipped with less information and processing competence, it does have some initial advantages. It is perhaps a welcome feature of the human mind that it is becoming more and more exacting in its demands to see theory or cognitive belief does rest on and/or corrigible by some objective or commonly sharable experiences. This intellectual demand underlies many intellectual enterprises of varying precision. Of them I shall briefly refer to only three, (a) philosophy, (b) scicnce, and (c) operations research, in that order.

(a) At the outset I shall readily concede that I do not believe in any sharp division between philosophy and science. The point can be illustrated both conceptually and historically. It is independently interesting and instructive to note how the generic discipline philosophy has in the course of time branched out into different disciplines like metaphysics and natural philosophy and yet how even now they are continuing to interact meaningfully and to their mutual advantage. My main concern in this paper though a little broader in its scope, has something to do with the history and the underlying rationale of the separation and the growing reunion of philosophy and science. It is not totally unrelated to the carrier of myth and magic, especially when we bear in mind the applied or the technological aspect of science.

It is not easy to define the concept of philosophy. It is an omnibus term which has been differently and often very loosely used to denote a wide range of subjects, from the pursuit of wisdom to the calmness of temper, from physics to metaphysics. There is nothing wrong in the inherent generalizing tendency of philosophy. But it entails certain problematic characteristics which should not be overlooked or underestimated. Called upon to organize different types and grades of experience and, if possible, to orderly systematize them, philosophy cannot help forming or borrowing and using general concepts. In order to be fair to a variety of facts of experience general concepts of wide comprehension run the risk of being inexact or, alternatively, losing their unitary character. It is true that is not necessarily the case. Inexactitude is excusable in the philosophical discourses which are primarily metaphysical in character. Many metaphysical concepts like "god", "spirit", and "life" are rather inexact in their claim and ascertainably indefinite in their comprehension. That does not necessarily render them useless. Perhaps these metaphysical concepts which should be better designated as categories, do serve the broad purpose of the rules of synthesis of kindred objects. When the concepts are very general and indefinite their uses become equally frequent and facile and consequently the *crucial* instance of their application which could be rightly regarded as genuine test cases turn out to be increasingly difficult to identify.

Concept is not container of things or a basket of objects. Its character can he grasped by two characteristics which are rather obvious and two others which are implicit. First, it is related to and perhaps a symbolic expression of a specific mental ability. A concept which has no perceptual correlate is a sort of unpredictable floatsam in mind and hardly enables us to speak of it and to use it purposefully. Secondly, it has an outward direction and reference which is publicly accessible in principle. Corresponding to every concept, there is an object or a set of objects which are meant by it but may not be necessarily perceived. There are objects meant by concept but not open to empirical inspection. Thirdly, this point is related to the first point. Mental capacity or its specific actualization is not an isolated phenomenon in the world of mind. Because of the conditions of spatial framework and temporal succession in mind, it occupies a definite position in relation to other concepts. The concepts taken as a whole tend to form a unity and intend or are directed to an outer world having a somewhat similar structural composition. This brings me to the fourth point. The objects in the world are also related in a definite manner which can be described by certain testable laws and predicted with more or less accuracy.

Certainly, all concepts of philosophy will not satisfy equally the above conditions. The concepts of metaphysical philosophy are formed and used primarily to describe the general features or structural properties of reality. In the quest for adequacy often these concepts turn out to be very ambitious in their scope, and at some stage or other it becomes difficult for one to decide where exactly to use them and where not to. In the absence of something inherent in the concepts suggesting the areas and limits of their application or somehow indicating the line of distinction between their use and misuse, it again becomes very difficult for one to stop the misuse of concepts. Further, once the line of distinction between use and misuse is erased or becomes practically unrecognizable, multiplication of concepts and of addressless ontological population prove unavoidable. The use of concepts is somehow related to some empirical posits. But the promise and power of this relation are contained in and underlying the mental capacity of forming and using the concept itself. The formation, use, and transformation of concepts are not a unilateral affair but of necessity related to and occasioned by some sets of appropriate empirical posits or objects. The appropriateness is, however, a question of practice or test. If the footloose and addressless concepts are uncritically formed and "used", the gap between metaphysical philosophy and natural philosophy would persist, if not widen. Whether some metaphysical concepts, despite their initial inexactitude, can or cannot approximate the relative exactitude found in scientific disciplines depends, among other things, on the four conditions, individual and structural, enumerated above.

When philosophers like Kant refuse to accept the strictures against metaphysics and propose to formulate it in a scientific manner their optimism seems to have been prompted by the perception that the gap between metaphysical philosophy and natural philosophy was largely due to the overstatement of the case of the latter and the understatement of that of the former.

(b) One does not know where exactly philosophy ends and science begins or, as some people would like to put it, where science ends and philosophy begins. Frankly speaking, as I have already said before, I do not know how to prevent completely the overlap between these two disciplines. In every scientific discourse one finds certain concepts and statements without which scientific objects cannot be meaningfully interpreted and related. Even the very object-character of what are called observables cannot be identified without some fairly exact concepts. The world of experience as rediscovered in that of theories is found to exhibit three distinct but related layers: (i) observables, (ii) objects, and (iii) theoretical entities.

Science has two main objectives: one, like all other factual disciplines, it aims at providing knowledge of reality; two, on the basis of that knowledge it wants to explain, predict and, wherever possible, control the endlessly complex objects of experience. As I have mentioned earlier, all objects, though in principle observable, are not actually and directly observed. For purposeful scientific observation we need concepts and, what is more, laws which can relate and interpret those concepts. Observables are scientifically systematized in terms of conceptualization and empirical generalization. In terms of laws of universal or statistical form, the relatively lower level generalizations seek to establish connection only among the directly observable aspects of objects. 'Fire burns", "river flows downstream", "cork floats on water and stone sinks in it" and the like are everyday scientific generalizations. More precise quantitative laws as Galileo's and Kepler's enable us to express a regular connection among directly observable phenomena and are themselves open to definite predictive test.

At a higher level of science we come across general statements containing such terms as electric, magnetic, and gravitational fields, atoms, electrons and a variety of sub-atomic particles. In the higher level psychological theories we encounter general statements which make use of such terms as ego, super-ego, self, subliminal self, id, libido, fixation, auto-suggestion and many others not directly observable entities. Vocabulary of empirical science contains two types of extra-logical elements, *observational* and *theoretical.* But the scientist is obliged to show that observational terms and theoretical terms are "somehow" related by laws and theories which lend themselves to predictive test and retrodictive use.

This "somehow" raises a lot of dust and difficulties, obscuring the relation between the theories and the observables. It is well-known that in between observables and theories there is something else, relational linkage, rule of interpretation, or whatever we call it. It is in this area of difficulties that one hears of the paradox of theorizing or of the theoretician's dilemma, which, in brief, is somewhat like this. If the concepts and principles of a theory can by themselves establish definite connections among the observables then the so-called third element is redundant and could be dispensed with; and if, for some reason or other, they cannot definitely connect the observable phenomena, then they are dispensable *ab initio.* In any case, it is concluded, the concepts and principles of 'theories are unnecessary. Obviously, this is an extreme position which is not likely to be espoused by any discerning thinker, and methodologist. If from a scientific world-view theoretical components could be completely eliminated, it might have been a matter of glee for the common man unequipped with complex and abstract theories. But the outcome of theoryless science based on pure observation, if any, would have cast gloom over the world of knowledge.

(c) Operationalism, an extreme form of practical verificationism, is perhaps the most instructive reaction against the theoretician's dilemma. The operationalist aims at linking scientific concepts and principles to experimental procedures and practical tests, eliminating thereby the operationally undefinable terms. The formulation of operationalism may be something new and attributed to scientific philosophers like P. W. Bridgman and B. F.

> Skinner. But it was already implicit in the working practice of scientists for a long time. Even the magician-scientist of the mythical world did perceive the necessity of putting his ideas to the test of practice (of various rites and rituals).

The programme of the operationalist is rather radical in its profession and ambitious in its scope. First, the claim that scientific concepts as such, i. e., irrespective of their theoretical context and system-affiliation, are operationally definable can hardly be worked out. Secondly, the apprehension that in the absence of well-defined operations the claims of competing scientific theories, different systems of geometrical measurements, for example, cannot be rationally settled, is prereflective and unfounded. *Per contra,* in order to determine whether certain operations are well-defined or not, we do need and use certain theories which do not lend themselves to operational definition. If, for instance, space is not *assumed* (preoperational assumption) to have an intrinsic metric of its own, then the choice and rejection of alternative systems of geometry or methods of measurement turn out to be merely a matter of convention and convenience and which in turn are bound to be *ad hoc* in character. In the most "glorious" days of operationalism it was maintained that the application of every scientific concept and the meaning of every scientific proposition depend upon their definability or otherwise in terms of performable physical operations. Besides the point already made to the effect that useful scientific concepts do not as a rule lend themselves to exhaustive definition, it can also be pointed out that the precise character of the defining operations is itself dependent on some partly or directly non-operational concepts.

The critique of operationalism in course of time persuaded the operationalist of the futility of chasing for exhaustive operational definition of each scientific concept. The monopoly of experimental operations and practical tests gradually yielded considerable place to "paper and pencil operations", mathematical and logical manoeuvrings, *indirectly* connecting concepts with instrumental operations. Finally, it has been admitted by the operationalist that operationalism is to be taken as a programme for eliminating inexact concepts and imprecise theories. No serious thinker is likely to raise any objection against this sound programme and sober position.

IV
THEORY AND PRACTICAL VALIDATION

I shall here use the term practical validation in a wider sense, including test, use and application. Broadly speaking, there are two types of views on validation, self-validation and other-validation.

(a) The self-validationist holds that theories are valid on their own and that it is only their invalidity which is due to something else, "other", or alien facts. The other-validationist opposes this view and contends that both for their validation and invalidation, theories depend on the alien or external facts. These views on validation and invalidation need some clarifications. If we say, as some people do, theories, if valid, are intrinsically so, and, if invalid, are because of some extrinsic considerations or facts, we assume a certain relation between theories and facts or objects of experience. Theories are called for to relate and organize objects of experience, predict them and, in some cases, to control them. Had objects and concepts designating them been inherent in theories, the question of validation of the latter in relation to the former would not have arisen at all. It would be too much to assume that theories are not only constitutive of the concerned objects but also valid *a priori* irrespective of the empirical relation which may develop or emerge in between them. This assumption is incompatible with my proposed, and what I think also correct, answer to the fundamental epistemological question, "do concepts *constitute* objects?"

Ontological independence of objects is a basic tenet of realism. And I for one do not see any good reason to question it. The questioning of it implies, among other things, the denial of the very basis of formation and transformation of concepts. Concepts are not there in an accomplished or a finished-for-ever form and our attitude towards them cannot be one of "accept or reject" them. If the character and career of concepts are themselves subject to the condition of social space and historical time, it is very difficult to hold that the objects "constituted" by the concepts can ever defy or bate the latter. In other words, in that case we have to believe that objects necessarily conform to concepts. Obviously, that is not our experience. We often construct or postulate concepts in order to identify certain objects which are not yet clearly experienced but in support of the existence of which we have independent and indirect empirical evidence. Even equipped with appropriate concepts, we at times can only delimit the area of possible objects and without being actually able to identify the objects themselves.

This point may be further buttressed by extending the argument from concept-formation to theory-construction. If the relation between theory and the area of its objects or of possible application could be known a priori and validly, interestingly enough, the task of theory-construction itself would have become somewhat easy, if not superfluous. We need theories not only to understand and explain but also to search, predict and control the concerned area of objects. Had theories been *a priori* constitutive of objects together with their interrelations, all explanation and prediction would in that case have been merely a matter of deduction. Further, if predictions turn out to be false or inexact, then only the composition of initial conditions has to be called in question, leaving the theories themselves completely untouched, for, we are told, they are valid a priori. The argument might be applied to the case of perception as well. If the rule of perceptual synthesis is deemed to be constitutive of the content of perception, then the possibility of unruly perception is straightaway ruled out. But all these hypothetical considerations are at variance with our experience. The case of a priori validationist rests on questionable assumptions which are unfair to the facts of experience.

(b) It is not easy to decide whether we rely more on theory or on practice in our quest for knowledge. Nor is it perhaps possible to offer any uniformly correct generalization on the matter. That we need both theory and practice for

> knowledge is well-accepted and almost trivially true. The interesting point to be noted is that in the disciplines where theories are said to be primarily important, the question of practice is not altogether ignored but only relegated to a secondary position. Similarly, one might point out that the disciplines in which practice is highlighted are not at all without theoretical foundation or presupposition. Sometimes, theories are trivially true and so well-known that they do not prominently figure at all in the open discourse and our attention is concentrated almost exclusively on practical questions. But a plain truth, and which is very fundamental in character, should be borne in mind: practice without theory is not at all possible and, theory without practice, though possible, is purposeless.

It has been rightly said that theories are like nets cast to catch (or construct) objects. Like all metaphors the net metaphor of theory has its limitations. The objects caught in the net of theory are not brought together for the first time by the spread and composition of the net and have definite relations between themselves independently of their position in the network of theory. The purpose of theory, among other things, is organization and systematization of an independently and definitely related (set of) objects.

Myths, like theories, are also designed and intended to organize and systematize objects of experience. But, as we have noted before, objects may be identified with varying degrees of definiteness. The concepts deployed for the purpose also vary in degree of precision. Besides, organization may be rigorous and categorical as in axiomatizable disciplines or loose and provisional. The character of system depends on the satisfaction of some well-known formal conditions and, what is more important from the practical point of view, simplicity and usefulness of the rules of interpretation linking the concepts and theories of the system with the intended domain of objects.

Practice is a generic term comprising various criteriological operations like verification, falsification, application, correction, precisification and so on. Volumes have been written on these criteria and their metamorphoses. The point of our immediate interest is that all these criteria are expressive of a common reaction against the deficiency of the theory of *a priori validation*.

(i) To ascertain and to be satisfied whether a particular theory is valid or not, one cannot exclusively rely on what is said to be pure reason uncontaminated by experience or self-certifying intuition.
(ii) The words like *ascertainment* and *satisfaction* must not be taken in personal and psychological sense. These two positions, (i) and (ii), taken together raise a question: if the operation of validation has to be completely impersonalized, then there must be a cut-off point between one's experience, on the one hand, and *what* is experienced and how it is so, on the other. Ordinarily speaking, and perhaps philosophically too, the concept of practice makes sense only in the context of some or other person. Practice is always that of a person or of a set of persons. Not that some philosophers of metaphysical persuasion have not spoken of *experience as such* without referring to some definite person or persons. Conditions of experience, including the subjective factors of the concerned persons, being different and varying as they are, explain to a great extent the lack of strict uniformity in the results of practice in the experiences of the self-same objects. If, in the quest for strict uniformity we resort to rigorous quantitative methods of logic and mathematics, we are often, not necessarliy, landed with paradoxical results. Certain concepts like equivalence, for example, are extremely difficult to define in an accepted manner. Consequently, we find it quantitatively very difficult to assess and compare the strength of two alternative hypotheses in relation to an identical set of verifying or confirming evidences.

Practical verification of universal and open-textured theories raises equally serious difficulties. Practical verification of a limited number of cases (or areas of objects) fails to provide any conclusive proof of the correctness of the theories in question. Probabilistic extrapolation of the results of *verified* cases beyond the empirically defined limits has its own inherent problems, viz. whether Nature is uniform and whether the variety of the kinds (multiform) of natural objects is limited, and the like. Little reflection is called for to show the unverifiable character of the metaphysical assumptions underlying the referred to probabilistic extrapolation.

Besides, the dispositional or propensity properties of the cases or objects under consideration pose further difficulty for the verificationist. To avoid the difficulty he liberalizes the criterion of *verification* and opts for *verifiability* which, to his understanding, can take care of the issues raised by universality, open-texturedness, dispositional or propensity character. Unless one retreats from the very commitment of empiricism, i.e. the requirement of sense-experience at some point or other, one can hardly conceive even of the *verifiability* of the universal and open-textured theories. If disposition and propensity are taken to be physical properties, undoubtedly the empiricist can take some comfort from the fact. Even then the problems raised by universality and open-texturedness continue to bedevil the empiricist view of practice of universal truths or theories. I must confess I see no compelling reason to doubt or deny more or less dispositional character of every theoretical statement. Even the so-called purely existential statement seems to have an element of dispositionality in it. The so-called physical character of dispositionality can hardly contradict its depth dimension, i.e. multi-dimensional and transcendent character.

The difficulties associated with the concept of *verifiability* obliged the scientific theorists to find out and formulate another and less questionable criterion of correctness. The result was the concept of *falsifiability*. Whether falsifiability was originally in tended to be a criterion of demarcation between science and metaphysics or that of empirical meaningfulness is not free from controversy. Historically perhaps it is true that this criterion was thought of primarily to draw a line of demarcation between what is scientific and what is pseudo-scientific or metaphysical in the bad sense. Nobody denies the importance and necessity of finding and formulating precisely the criterion of empirical meaningfulness. But certainly this cannot be the sole or even the main preoccupation of any scientific theorist. Understanding or misunderstanding of the intended meaning of a theory certainly makes the difference to one's approach to test it. And therefore, there is no denying the fact that one must make best use of the relevant rules of syntax and semantics for the ascertainment of the theory before its practical test is undertaken or devised. But at any particular point of time one's ability to understand the meaning of a theory is governed by certain subjective and objective factors which make it difficult for anyone to assert with definiteness whether the concerned one's understanding is absolutely definite. This proposition, though sounds very general, is basically sound and follows from the basic tenet of anthropological rationalism and epistemological fallibilism. This proposition is also evident from the history of many scientific concepts and theories. The changing and growing characters of knowledge makes it imperative for us to find out the correctness of the truth-claim of empirical theories. The question of truth is more fundamental than that of meaning. We are now primarily engaged with the issues concerning practical tests.

The falsificationist's way of testing the truth of an empirical theory is very simple. He does not try to prove, support, confirm or justify a theory. He proposes to move just the other way round. The scientific man must try to disconfirm a theory by finding out contradicting examples. A universal proposition may be contradicted by a negative existential one. What is most important is the actual discovery of disconfirming instances and not the mere psychological intention to falsify a theory. Discovery of a single falsifying case partially damages a theory but does not then totally destroy it. Disconfirming cases define and delimit and again redefine and redelimit a theory. Attempts to falsify a theory obviously do not succeed in all cases. The relevant cases which fail to falsify are counted as corroborating instances of the theory. Corroboration is not confirmation. But while the former embodies an attempt to justify a theory, the latter is a proof of the fact that theory has survived the test and in the process provisionally established its truth-claim. The whole argument of the falsificationist of course presupposes that the falsifiers as expressed in existential propositions are themselves basically true. If the so-called basic propositions used to disestablish general theories are themselves equally or simultaneously liable to disconfirmation or disestablishment, their claim as falsifier cannot be entertained as serious. Here arises a related question. Can we defensively and critically believe in the existence of strictly existential propositions which are really basic in the sense that they are unquestionable for ever? I think that every existential proposition is more or less dispositional and therefore has a more or less "transcendental" and "multi-dimensional" character. Generality and existentiality are to my mind a question of degree. No existence can be absolutely pinpointed or pictured in a proposition. In whatever form or structure we capture or picture it, it has something in it which exceeds the bounds of the form or structure in question. A well-established scientific theory is in fact only damaged by relatively lower-level disconfirming instances and the nature of damage and the damaging findings suggest the way of correcting and improving the damaged theory in question. A theory which could possibly be completely destroyed by counter-examples is very unlikely to find serious propounders and defenders.

The metamorphoses of *verifiability* and *testability* have sobered down the scientific theorist and made him realize the advisability of finding out and formulating such modest concepts of practical test as corrigibility, precisifiability and empirical applicability. Every empirical theory is subject to correction and which could be achieved by applying it to appropriate cases. The appropriateness of the cases is *suggested* by the very nature and meaning of the theory. The said suggestion may be general, vague or rather imprecise. It is in and through application that theory is corrected and precisified. Correction and precisification presuppose *independent existence of the objects,* cases or areas whereto the theory is applied. To put it otherwise: if what possibly would correct and precisify a theory is not only suggested but also definitely "constituted" by the theory itself, it cannot do what it is called upon to.

Independence of objects has another aspect. Objects are not only independent of theories but also *relatively* independent of each other. In spite of what we call its multi-dimensional and transcendent character, an object does not *exactly* represent an infinite number of objects. Any general theory about a class of objects has to be carefully formulated and defended. For what is designed to be true of an infinite or at any rate indefinite number of objects may not in practice turn out to be so. Even when a theory *explains* a wide range of objects it does not explain each of them equally. To say that explanation by definition is general and true of a class of objects does not necessarily mean that it renders all concerned objects *equally* intelligible or that it is true of all objects equally. That is one of the reasons why explanation often turns out to be *explanation-sketch* rather than full-fledged explanation. Again that is why at least some *individual* objects covered by a theory have a say in the matter of correction and precisification of the theory itself.

Ontologically, what all this means is this. The world of objects which enables us to test a theory by practice seems to be a totality of related particular; and not an intrinsically unified whole. Patterns and sub-patterns of the relations between the particulars may be repeated but not necessarily in an identical manner and indefinitely.

This brings me to my next point. All theories are not equally practical as they are not intended to be so. Broadly speaking, there are two types of theories, namely, (a) those which are primarily intended to be of practical use, and (b) those which are primarily intended to be of theoretical use. Mathematics, metaphysics and theoretical physics are some of the disciplines marked by primacy of theory. No one would deny their practical relevance and, at the same time, perhaps one can easily indicate the difference between their practical relevance, on the one hand, and that of the disciplines like medicine and engineering, on the other. Medical and engineering sciences are primarily intended to be of practical use. When I propose to classify some disciplines in terms of *primacy of theory* and the rest in terms of *primacy of practice,* I for one do not forget for a moment that this classification is not exclusive or hard and fast. For there are disciplines which could be viewed both from this end and also from that end. Aerodynamics, for example, could be conceived both as a theoretical discipline and also as a practical one. I am sure one could think of other examples.

The relation between the disciplines which are primarily theoretical and those which are primarily practical may be understood in a clearer manner by trying to answer the question: whether theory precedes practice or practice precedes theory? This question seems to me purely conceptual. True, conceptualization also serves in a sense a practical purpose. But the operation of conceptualization could be carried on with only a minimal practical involvement or awareness. Obviously, the more important and interesting aspect of the relation between the precedence of theory over practice and that of practice over theory has to be understood from a practical point of view. Theory and practice are practically inseparable movements of the same process of exploration and organization. The questions of primacy and precedence also could be better understood from a practical point of view. Every act of theorization has a practical motivation behind it. And this motivation may be articulate or inarticulate. Theory guides us and leads us to action; and action in this context has to be taken in a very wide sense including exploration, understanding and cognition. Practice corroborates or corrects, in most cases both corroborates and corrects the lead and guidance provided by theory.

The question of primacy of theory or that of practice is partly influenced by one's view regarding the relation between the universals and particulars. When theories are formulated on the assumption that universals *qua* universals do exist and that the main purpose of theory is to provide structural description of those universal features of the world, the difference and refinement suggested by the particulars or the experience of the same are bound to be more or less neglected or relegated to a place of secondary importance. The world-picture found in mythical and speculative metaphysical theories is not constructed piece by piece or brick by brick. In other words, mythical and metaphysical world-views reveal no special perception of particulars. Whenever particulars are allowed or tolerated in speculative world-views, those are portrayed as ascertainably unrelated and yet predictable examples of unspecified universals. Diminution of the value of particulars lead to a sort of *a priorism.* "Accidents" and "miracles" are ordinarily unknown visitors to these types of world-views. But they are sought to be explained in terms of still higher order and universal principles. The more a theory is marked by primacy of practice, the more it is found to be a respector of particulars. Primary practical motivation of a theory obliges it to address itself to a specified area (of objects) marked by some definite particular characteristics.

When the theorist knows that his theory has to establish its bonafides in terms of its application to concrete details and specific objects, his theory increasingly assumes a practical orientation and realistic character.

V
IDEOLOGY AND PRAXIOLOGY : SCIENCE OF ACTION

Views about the nature of idea are numerous. At the moment I shall refer only to two of them. First, idea has often been understood as something given, independent and permanent. One may fail to grasp it for some reasons or other, mainly subjective but not necessarily so. One's failure to grasp the static and independent nature of idea has nothing to do with the nature of idea itself. Elements of Platonism are evidently associated with this concept of idea. But there is another concept of idea which takes it as a programme or scheme of action. According to this view, idea as idea is not self-contained or an end in itself. Idea is pregnant with a sort of will-to-act. The volitional component of idea takes away its static character and imparts to it or rather, one might say, reveals its other dimension. Idea has in it something which enables it to transform and exceed itself. Idea may be characterized as scheme of action, and action as transformed idea. In this connection some thinkers refer to the concept of intention, intention as an inherent element of idea, leading to some objects or action. Object or intention is taken to be the meaning of idea. Meaningful ideas are self-exceeding, self-transforming, objectward and action-oriented. Idea is arrested action.

The assumption underlyirg the second concept of idea, though inarticulate, is unmistakably psychological. One might even say that there is an element of subjectivity in it. If the Platonic elements associated with the first concept of idea are disregarded or, even otherwise if we think of idea in a commonsense manner, we will be inclined to accept the view that the human mind is the natural locus of (and also perhaps impetus to) idea. Let us take note of the fact that of late very sophisticated attempts have been made to define mind in some absolutely non-subjective manner and in terms of some empirically evident and objective performances or acts.

The question whether this type of definitions of mind and the associated concepts of ideas are consistent with the notions of human responsibility and dignity is not free from controversy, and which appears to me very interesting.

Idea may or may not be accepted as scheme of action. Even if it is, that does not necessarily mean that there is a value component in it. Scheme of idea may be appropriate or useful without being good. Again, ethical concepts like good and bad may be understood in a naturalistic manner, totally disregarding what is called its intrinsic value character. Some thinkers are found to have highlighted the concept of ideology as distinguished from that of idea (as scheme of action) which may or may not be accepted from a valuational point of view. The possibility of ideology itself has been questioned on naturalistic grounds.

Naturalism itself may be of two types, empiricist and realist. The empiricist critic's main point against ideology is that it involves abstraction from and trancendence beyond an experience. According to him, the safest and the surest course of action demands of us to be as close as possible to our own experience. But ideology is bound to be somewhat abstract in character and therefore more or less alien to a given set of problems in a given situation which we are called upon to tackle or solve by it. I admit that there is some substance in the pro-empiricist criticism against the possibility of ideology. The concept necessary to organize experience and meant for application to objects cannot be derived from either experience or objects as such. For, organization of experience and identification of objects presuppose the concepts and the competence to use them according to appropriate rules. Similarly, human action presupposes some scheme of action, however inarticulate or provisional that might be. However, if the concepts are alien to objects or intended experiences, and schemes of action turn out to be useless for the purpose of realizing the given aims of action, concepts and schemes are easily dispensable. The question is whether ideology belongs to the category of inapplicable concepts and useless schemes. This question may be given a logico-epistemological form: does ideology owe its origin to induction or extrapolation from experiences and actions as such? Is it not the case that human action, as distinguished from physical action, presupposes some idea or scheme? Further, is it not the case that object to be recognized as object, experience to be organized as meaningful and identifiable entity, concepts are indispensable? The relation between concept and object and that between scheme of action and action cannot be rationally understood without some appropriate rules. These rules may not be fixed and inflexible, but their character, composition and function must be more or less definite and anticipatable. Otherwise, different and even conflicting concepts could be deemed to be applicable to the same set of objects; and, to put the matter from the other end, the same set of objects might equally confirm or disconfirm different and even conflicting concepts. Similarly, one might point out that if schemes are irrelevant to the success of action, then any and every scheme might lead to equally successful action. Certainly the world of concepts, objects and actions is not so hopelessly irrational and chaotic.

The realist critique of ideology is based on a reductionist account of value, which may or may not be Marxist in its inspiration. The very possibility of ideology has been contested. It has been argued that, rightly understood, ideology is a value-neutral concept. When unrealistic subjective and emotive elements are injected *ab extra* into ideology, it becomes uncoordinated with the rest of its situational components, inwardly somewhat disjointed, and practically unworkable. An ideology is not worth its name if it does not work. Workable ideology is nothing but a realistically drawn up scheme or strategy of action. Thus the realistic critique of ideology could be given a less *reductionist* look and a definite gradualist interpretation. Marx for one would not agree to reduce ideology into a mere strategy of action and levelling down its value-component to the material conditions of life. In between the two he draws a dialectical line of demarcation. There is a level-distance, perhaps even a categorical distinction, between the material conditions of ideology and its own character which has a super-structural and relative autonomy of its own, although some of its elements, including the valuational ones, could be dialectically, traced back to the basic material conditions of life.

The realist view of ideology, whether it is reductionist or gradualist, need not necessarily be practice-oriented. In other words, ideology and praxiology are not coterminous concepts, meaning the same sort of stuff. One may easily conceive of ideology exclusively as a dispositional or an attitudinal matter. A theorist or researcher, for instance, certainly can entertain an ideological disposition and develop some related attitudes without practically involving himself in what the said ideology might suggest as its practical corollary. What I mean is this. The *autonomy* of ideological dispositions and attitudes should not be ruled out *a priori*. While I do agree that psychologically it is difficult to entertain an ideology delinking it completely from what is ordinarily taken to be its practical corollary. Psychological sequence or "corollary" is no proof of existence of a *logical* connection between ideology and practice. There are some "incurable" theorists and "irresponsible" ideologists who do not practise what they are psychologically committed to. Their conduct suggests that there is a practical cutoff point between ideology and praxiology.

The above distinction drawn between ideology and praxiology, though I concede as a point of practical possibility, appears to me very tenuous and I wonder whether any coherent theory of mind can reliably indicate the supposed cut-off point between theoretical dispositions and attitudes, on the one hand, and practical ones, on the other. For analytic purpose we may speak of the distinction between thought and action, theory and practice, but as a matter of fact they overlap and interpenetrate. The hang-over of Faculty Psychology should not make us believe in a non-functional distinction between non-existent compartments of mental life. The supposed distinction between theory and practice and also that between ideology and praxiology is at times over-emphasized also for an obscure ontological reason. Dispositions, attitudes and thought are clubbed together and designated as category of reality called mental and, one argues, it is totally different from the phenomena regarded as physical (and perhaps physiological). The concepts like action and practice are totally different from those of thought and attitude. The defender of this sort of ontological dualist would like to maintain that *motion* and *emotion,* for example, belong to entirely separate categories, despite their semantic affinity. Given this formulation of the view, it contradicts and throws overboard even the minimally accepted core of the James-Lange theory of emotion.

Positively speaking, ideology, to my mind, is a composite

notion of things and beings which is consciously accepted and with an intention to work out, apply, implement or practise. I am not trying here to give a definition of ideology. It is an attempt to describe the real state of affairs to which the term may appropriately be applied. The point I propose to highlight is that ideology is generally practice-oriented. The cases, if any, where ideology is accepted or professed but not practised, should be regarded an exception and not rule. Further, I am inclined to believe, as I have already said before, the supposed distinction between theory and practice and the concept of practice-neutral ideology is an untenable legacy of Faculty Psychology, theoretical creatures of an indefensible ontology and the result of confusion between what is analytically possible and what is the real state of affairs.

I know my view of ideology, basis of valued action, is inconsistent with both positivist and materialist formulations of ideology and praxiology. Let me have a close look here at the materialist and interesting account of praxiology. According to this view, praxiology should be taken as a technical-procedural science giving a systematic account of (a) general guidelines, and (b) specific rules of action and response. The aim of praxiology is to lay down and improve the conditions and specifications of *quality* action and performance. Some of the conditions enabling us to improve the quality of our action and excellence of performance are general and somewhat flexible in character. Environmental conditions, immediate and mediate, laws of the land, social forms (I am deliberately avoiding the use of the term *norm* at this stage), and extant rules of the place of work are some of the accepted general guidelines which influence and are therefore relevant to the quality of human actions. To ensure the performance of quality actions, general conditions are sought to be *regulated, controlled, standardized* and *stabilized* but some of the conditions of quality action have to be further specified. This exercise or *specification* involves extensive use of *quantitative and statistical methods and techniques*. Physical fitness, adequacy, quickness and correctness of responses, span of performance-time, precision, accuracy or correctness of the performance of the task assigned together with the time span are some of the specific conditions of quality action which could be more or less quantified.

The main idea behind the identification and, as far as possible, implementation/realization of the conditions of quality action is to ensure and improve what may be called human *efficiency.* The concept of efficiency occupies an important position in the materialist praxiology. Efficiency, attained or imparted and ensured, leaves its imprint on the person, his product and the involved process of production. Efficiency is an enabling, adjustmental and improvemental disposition. Rightly understood and practised, it improves the quality of both *individual* thought and action and *collective* thought and action.

Besides quality and efficiency, other two key concepts of materialist praxiology are *information* and *decision.* Both quality action and thought are very much dependent on and conditioned by necessary information and appropriate decisions. It would be only fair to mention here incidentally that the elements and techniques of information collection, codification, computerization and communication are not peculiar to the materialist concepts of quality and efficiency. Being parts, as they are, of mathematical and engineering disciplines, they have their general use and applications cutting across the border-lines of different ideologies and worldviews. Perhaps the same cannot be said of decision (theory). I say "perhaps" because sophisticated attempts have been made to treat *decision* as an almost mathematical function of correctly communicated and understood information. Again I am using the qualifying term "almost" because I am not yet philosophically persuaded of the adequacy of mathematical laws and techniques of information (theory) and decision (theory).

To perform quality action in a given situation precisely which information are relevant and which of their implications have to be taken into account are not easy to decide and define. Secondly, the very formation of information is subject to the correctness of the choice of the concerned concepts and their adequacy. Stimulus-response operations do not automatically generate information. Concepts are absolutely called for forming, encoding or symbolizing and transmitting information. Since a concept or a set of concepts is used for the purpose of forming and encoding an endless number of relevant information, there is bound to remain a permanent, may be marked by fluctuations, epistemological gap between the two necessary components what may be called an informational whole. Thirdly, in the process of codification or symbolization of information materials we have to undertake one or more levels of translational/transformational activities, e.g. from natural language to artificial language, from quality to quantity, depending on the materials and forms of information, readable and/or intelligible, audio-visual, purely auditory, purely visual, and tactile. Fourthly, the problem of uncertainty systematically bedevils the rules of interpretation of symbolized and encoded information. In spite of rigorous definition and standardization of the rules of interpretation for decoding, the human or epistemic factor has its say. Codification and communication of nonsense syllables, phrases and sentences prove always a little different. Statistically speaking, elements of errors creep in an increasing manner. Meaningfulness or otherwise of information materials makes always a difference to the process of successful communication. It is in this context that one has to recall the crucial role played by concepts in converting information materials to actual information. Fifthly, the process of communication and transmission is disturbed and distorted not only by the human factors but also by the mechanical, electrical and thermal ones. Views are brightened, dulled or darkened; similarly, voice pitch made louder or lower; and the various other forms of transformation and/or transmission are mechanically affected. Unintended (weather) disturbance and (thermal) resistance are also occasionally experienced, which I admit, may be controlled to some extent, but not beyond a limit.

When I speak of the deficiency and defects of mathematical-mechanical information theory, its ingredients and implications, I do not propose by any chance to dispute their immense usefulness. The only point I want to make out is that ideology or praxiology cannot be derived entirely from or on the basis of correctly processed and communicated information. Besides the limits of information-based praxiology indicated above, I have some positive considerations why it should not be construed in a secular or value-neutral manner.

The most important consideration suggesting the valued character of praxiology (and for that ideology too) is to be found in the concept of decision. Here, again, right at the beginning I would like to draw a distinction between the techniques of decision-making and the conceptual components of decision itself. Given a situation together with its constraints, certain aims, interests and choices, it is always possible to calculate what choice, if made, would be the right choice. Other things apart, one should carefully observe the use of *right* in this context. *Correct* might replace *right* here. Bearing in mind the value-neutral sense of the word *correct* perhaps, one might say, the use of it rather than that of *right* would be advisable. By deductive reasoning one could be formally assured of a *unique* choice or a set of choices within a *definite* minimax range. But the point is that if the reasoning pertains to a type of choice where factual considerations are relevant, the mere observance of the formal rules of deductive reasoning cannot *guarantee* the correctness of the choice. For the factual premise of deductive reasoning is implicitly dependent on some questionable standards by which the types of facts and their relevance and scope are determined. Besides, the "facts" of the situation are accepted or questioned as a result of implicit analysis and interpretation; and therefore the rules of concerned analysis and interpretation are also subject to review and scrutiny. Finally, perhaps it could be shown by analysis of the "facts" that their *acceptance* is a result of *evaluation* and therefore should better be called an "objectively limited" *choice.*

If the forms and rules of deductive reasoning provide no guarantee to the correctness of decision, those of inductive reasoning are even more inadequate for the purpose. From some factual premises by inductive extrapolation we cannot get to any *necessary* truth. Obviously in making a correct decision we do not use primary induction which enables us to get to universal propositions by consideration of instance falling under them. Nor do we use secondary induction which is meant to establish hypothesis which relates to observables of a certain kind to some other things which are not observable. It is well-known that support or justification of induction cannot be absolutely perfect and is bound to be probabilistic in character. Given these inherent difficulties of inductive reasoning, one would not be inclined to accept it as a safe guide to what might be called correct decision. The case of correctness of a decision is patently more a matter of application of a general law (or a set of laws) than that of establishing the same.

To realize sufficiently the complexity of the method involved in correct decision-making several things have to be borne in mind, viz. (i) factual statements, (ii) competing rules of interpretation of facts, (iii) process of reasoning, (iv) statements of the foreseeable consequences of the decision reached, and (v) the purposes and interests contemplated by the rules of interpretation. The problem is, even if all the elements of correct decision-making are duly and sufficiently attended to, one cannot be sure that the decision taken is in fact the *right*. In other words, I come back to the distinction drawn between *correct* and *right* and to which I have already referred.

Right decision has a *genuine* choice element in it. The thought or theoretical element which accompanies and precedes practice or action is inseparably connected with the psycho-somatic process of deciding and trying. And these processes cannot be predicated purely on the basis of given information and aims. For the relevance-range of the information and the given aim undergo changes. Because of *trying, reviewing* and *re-evaluation* of the aim, certain forms of practice are inhibited, arrested or even controlled. And all these presuppose *freedom of will*. Whether certain conceivable forms of practice can at all be tried depends partly also on the agent's re-appraisal of the situation. Every agent's appraisal and re-appraisal of a "given" (or "his") situation is partly coloured or (feedback) influenced by his understanding of the aim and intensity of interest in and *commitment* to it. He *expects* some results to come off his *attempts* to realize the aim. The course of practice, the type of attempts made by him, is subject to *what happens to him*. What he can practise or do is related to what happens to him. The agent's *autonomy* of practice is not to be confused with the sovereignty often mistakenly attributed to his *self*.

Had a decision and the related course of practice been a function of a set of necessary and sufficient conditions, including information and strategies, the same could be safely predicted. Decision and practice can be inexactly *anticipated* but not exactly *predicted*. For the psycho-somatic processes of decision-making and executing (i.e. practice) are *open* to various influences, internal and external, naturalistic and evaluative. Consequently, theory and practice, thought and action, though theoretically distinguishable, are practically continuous and connected. The connection can be analytically followed or even guessed but, as I have said, cannot be safely predicted.

An adequate theory of theory and practice can hardly afford to neglect the fundamental ground of *unity of theory and practice,* thought and action. The concept of *person provides the ground for unity and continuity* between the two. Person is basically a biosocial concept. This or that individual person's choice of aim, choice of the course of practice and the rules concerning the both are more or less derived from or at any rate influenced by the society he lives in. I say "more or less", for the individual has always his inalienable freedom not to be completely influenced by what he is, has and what happens to him.

This freedom is evident also in the sphere of relation between idea and will, idea and practice. There is always an element of will in an idea which could be regarded as an overplus—something over and above what is indicated and circumscribed by idea. *Volition is not necessarily and neatly conceptualizable and exactly judgemental in character.*

Admittedly imperfect, *unity of theory and practice* is found not only in the life of individual persons but also in that of *society as a whole.* The unity of the two in person is not by itself a guarantee of the unity in the society as a whole. Unity in society involves something more than unity in the life of individual persons. The latter is necessary but not sufficient for the former. Once we accept the position that genuine ideology is possible and, what is more important, desirable, it is difficult for us to oppose individual's responsibility for bringing about some change in the existing social order and the initiative he might take towards that end. The responsibility and primary initiative of course lies with the individual. For he is the primary focus of perception and attention of social needs and provides the primary impetus for social action. The issues become very important when we bear in mind the theoretical justification of what I call volitional overplus. An individual's idea of an ideal society might and in fact often does transcend the bounds of contemporary good and just society. A creative or revolutionary individual can and will and try for something more than what is contained in the given idea of good and just society. In his bid to establish a revolutionary and/or creative society he might peacefully try to transform and, failing which, violently destroy a part of the existing social order. Revolution need not be necessarily creative to start with. It may take time to disclose the creative impulse and identity of an apparently destructive revolutionary enterprise.

On the basis of this argument almost every social order can be threatened and disturbed. Naturally one could ask for *the rationality of revolutionary theory, practice and their unity.* Rationality of a theory of revolution (and for that matter of any action) is to be sought in terms of its practical possibility to deliver the intended good aim or objective. Progress may also be indicated as a parameter of the rationality of a revolutionary theory. Internal coherence and situational compatability may he cited as other two parameters of the said rationality. The critic can always raise objection against what has been called *good* aim or objective. This raises some fundamental questions about the key ethical concepts like *good* and *just* (society). The less fundamental but very interesting questions pertains to what is to be called *progress.* Is progress a routine achievement of human efforts? Are almost all of us inching towards its goal, as a matter of course? Or, the critic may also ask, can revolutionary progress be achieved in a routinized way and as a matter of course? Does it not involve breaking away with the existing routines and normal courses? Admittedly we can replace one routine by another and one course by another, but the question is can we do away with all routines and all courses and yet make Revolutionary progress possible and attain a revolutionary goal? Revolution may be festival of the masses against the routines and courses of the classes, but that too has to be observed according to some rules and norms. The main difference between the revolutionary practice and the non-revolutionary one is to be found in the composition of authors and the structure of authority. Otherwise, as we have seen in many cases, in spite of their creative impulses, many revolutions degenerate into lawless orgy of the masses and end up in self-defeating anarchy. For even the sincere participants in revolutions could not grasp and apply the rules of initiating, leading and sustaining a revolution through a sequence of quickly changing situations. The genuine revolutionary has to first theoretically indicate (but need not neatly and completely work out) the rationality of a revolution in a given situation before he proposes to practice it in a disciplined manner and honestly expects others to join him in this creative adventure of existence.

REFERENCES

1. Brodbeck, May, *Readings in the Philosophy of the Social Sciences,* New York, 1968.
2. Chattopadhyaya, D.P., *Societies and Cultures,* Bharatiya Vidya Bhavan, Bombay, 1973.
3. Davidson, Donald and Patrick Suppes, *Decision-Making: An Experimental Approach,* California, 1975.
4. Gasparski, W . W. and Klemens Szaniawski, "Praxiology and Decision Theory", Warszwa, 1975.
5. Habermas, Jurgen, *Theory and Practice,* tr. John Viertel, Hienemarn, London, 1973.
6. Hampshire, Stuart, *Thought and Action,* Chatto and Windus, London, 1970.
7. Harre, Rom, (ed.) *Problems of Scientific Revolution,* Macmillan, London, 1975.
8. Held, Divid, *Introduction to Critical Theory: Horkheimer to Harbermas,* Univ. of California Press, Berkeley, 1980.
9. Korner, Stephen, *Experience and Theory,* London, 1966.
10. Levi-Strauss, Claude, *Savage Mind,* Weidenfeld and Nichlson, London, 1966.
11. Popper, K. R., *Conjectures and Refutations,* London, 1963.
12. Putnam, Hilary, *Reason, Truth and History,* Cambridge Univ. Press, 1981.
13. Radnitzky, G. and Anderson, G. (eds.), *Progress and Rationality in Science,* Reidel, Dordrecht, 1978.

11

SATYAGRAHA AND SOCIAL CHANGE

It is possible only for few persons to speak profound truths without theorizing and using jargon. Most of Mahatma Gandhi's influential concepts are extracts and concentrates of experience. Theorizing is not native to his nature. Jargons do not come to his mind at all. Yet his ideas and ideals are graphically clear whether we accept them or not. His illustrations and practice make them even more vivid to our mind. Gandhi does not claim to have any "set theory to go by".[1] He confesses to have not been able to work out "the science of *Satyagraha* in its entirety." Yet his views on the matter are clear enough to give us a correct idea of it and the connected considerations.

To highlight the positive content and meaning of the idea of *Satyagraha* it has been distinguished from the kindred idea of passive resistance. Satyagraha literally means the drawing force or attractive power of truth. Gandhi uses several synonyms of *Satyagraha*, viz. Truth-Force, Soul-Force and Love-Force. Neither passive resistance nor Satyagraha is an altogether new idea. He points out (1908) that the idea of passive resistance is as old as the human race and has been commonly practised in India long before it came into vogue in Europe. Bhishma and Vamadeva are said to have defended this doctrine as the inherent right of the common man against tyranny.[2] Mention has been made of different forms of passive resistance, *dharna, hartal* and *deshtyaga.* Sri Aurobindo has also defended (1907) passive resistance as a political strategy "for the vindication of national liberty or in order to substitute one form of Government for another, or to remove particular objectionable features in the existing system without any entire or radical alteration of the whole, or simply for the redressal of particular grievances." While Sri Aurobindo prefers passive resistance, bearing the contemporary conditions and preparedness in mind, he does not rule out *a priori* the necessity of armed revolt or aggressive resistance. He does not think that armed political movement is criminal or unjustifiable in all circumstances. Freedom is man's natural right. To defend it he can resort to use of arms. The right to resist the wrong was defended on the analogies of the English Dissenters and Hampden's refusal to pay ship-money. Referring to the long and bitter struggle between the Stuarts and British Parliament in the seventeenth century, C.R. Das defends passive resistance both on legal and moral grounds. Das's main argument is that law of the land has no sanction behind it unless it is based upon the consent of the people concerned.

The controversy whether the doctrine of passive resistance drew its inspiration from the Indian tradition or from the Euro-American history or, what is perhaps more plausible, from both seems to be rather trivial.[3] Conceptual and psychological aspects of the question are much more interesting and important. Human resistance under normal circumstances is found to be passive to start with. And it is also peaceful. Armed resistance requires a lot of provocations or a sustained provocation and necessary preparation. *Unwillingness, disinclination, hesitation, reluctance* and the like are the words appropriate to describe the initial phase of our mind leading to a state of resistance. Disapproval or dissent are more definite dispositions. All these words, including *passive resistance* itself, designate a negative state or disposition of mind indicating some or other value-judgements with reference to a definite aim and object.

Satyagraha is definitely a positive concept. It expresses a positive disposition of mind and is clearly indicative of mind's drive or inclination towards a valued object. Gandhi presupposes and relies upon a particular view of human identity to defend his view on Satyagraha. He says, it is a Soul-Force, a force welling out of the very nature of human soul. Unless human soul is enveloped and affected by some alien physical forces, it expresses and feels drawn towards what is true. He also speaks of Truth-Force, suggesting thereby that truth is not a mere "logical notion", and it has "substantive existence". Gandhi has confessed his pro-Vedantic inclination at many places. Truth is self-revealing. Soul is self-revealing. Truth is God. And the best expression of God in man is his soul. With some minor interpretations this part of his view is quite consistent with the Vedantic outlook. But when he speaks of Satyagraha as Love-Force, he sounds a bit Christian or *Vaishnav*. I am aware that it is not fair to assess a mind like Gandhi's in terms of concepts drawn from other persons and their systems. But for limited analytical and comparative purposes, perhaps it is interesting and even illuminating. What in fact Gandhi wants to highlight when he uses the concepts of truth, soul or love is that Satyagraha is not something physical or material, and is not prompted by any economic considerations. In other words, it is not occasioned or actuated by any alien principle or law. The fountain of truth is the soul itself. One might go even further and say, *Satya* (Truth) and *Atma* (Soul) are identical at bottom, although the former is primarily used in logico-epistemic context and the other in the metaphysical context.

Gandhi is aware that Satyagraha may be inconsistent with the law of the land at a particular time. He has developed a set of arguments to meet this rather legalistic objection. First, he points out that, although tradition is a very important thing in the life of a nation, its perpetuation or prolongation beyond a point of time cannot but inhibit freedom and creative genius of the people concerned. Repetition of history is not the lesson of history. The lesson of history is to change and enrich it by discoveries, inventions and innovations.[4] Gandhi finds no incompatibility between tradition and modernization. Secondly, he characterizes Satyagraha as God's law.[5] Man is essentially divine. Continuous self-expressiveness is intrinsic to God's and therefore man's nature. No law or set of laws can stand in the way of this natural or divine process of self-expression, and if it does, it has to be questioned and opposed. When man-made law proves inconsistent with God's law, one must follow the latter and reject the former. This is true both for the individual and also for the collective. Finally, Gandhi supports his contention by a rather strong argument. To tolerate tyranny and injustice is not only politically indefensible but also morally degrading.[6] Foreign rule violates the basic precept of democracy. A Government which governs without the consent or against the will of the people concerned is definitely anti-democratic.[7] Gandhi did not deny that there might be few good aspects of the British rule in India but he was not interested in removing only some particular objectionable features of the imperial system. He wanted to remove the system itself. He is very forthright on the point. He is for the wholesale rejection of the British rule for, in spite of its good aspects, it is like "the fabled snake with a brilliant jewel on its head, but which has fangs full of poison."[8]

The theological and metaphysical foundation of *Satyagraha* has wider implications. The Soul-Force is universal or universalizable at any rate. When soul realizes its inner essence, it finds itself in all souls and also all other souls in its own soul. A fully realized soul knows the essential identity between all living souls, and therefore the Soul-Force in its best form is universal. When this force is actually universalized, it can "revolutionize social ideals and do away with despotism and... militarism." The force of the despot or of the military dictator is primarily physical and material. Its success is bound to be short-lived. It evokes hatred. Permanent human relation can be established only on the basis of mutual love and respect. The aim of Satyagraha is to persuade and convert and not to coerce. The process of conversion touches the very core of human existence and awakens and activates the Soul-Force of man. The other way, the violent or coercive method, of achieving an end is squarely condemned by Gandhi.[9] Imperialism fails because it relies only on the armed forces. He concedes that every nation and every individual have the right, and it is their duty to rise against foreign rule, imperialism or an intolerable wrong. But he is definitely against "armed risings". For "they are a remedy worse than the disease sought to be cured."[10] Violent revolutions or armed risings are symptomatic of the spirit of revenge, impatience and anger. The *Satyagrahi* must not allow himself to be motivated by these negative and base forces of human nature. These forces are neither stable nor constructive.

Satyagrahis must satisfy some basic requirements of character.[11] He must be *honest.* He must be not only *disciplined* but also non-violent in thought and action. Finally, he must be prepared to *deny himself* all that he can possibly have, like wealth, family and other possessions. A word of caution. Gandhi does not encourage any ascetic bias. Suffering is necessary but has its "well-defined limits... and when the limit is reached, to prolong it would be... the height of folly."

Satyagraha has two aspects in it, individual and collective.[12]

Their root force being identical, the moral requirements of the Satyagrahi, whether he is an individual or it is a group, are also identical. Besides the basic character-traits of Satyagrahi indicated above, Gandhi has spoken of some other requirements as well. Satyagrahi must be *truthful,* truthful to his inner voice or conscience. He must be *faithful* to his ideal. His plan of action must be *thorough.* He must be *fearless* without being arrogant and develop humility without timidity. He must be *steadfast in following* his ideal in practice. He has to be *punctual* and *orderly* in his actions and movements. He should *know his own limitations* and be able to *restrain* himself whenever necessary. If something goes wrong, he should be morally strong enough to *own his responsibility,* and take upon himself the resulting blame.

It is obvious to anybody that Gandhi's Satyagrahi is an ideal type and it is not easy to find out a perfect and concrete example of it. But that is perhaps no strong argument against his view as such. He speaks very clearly about the necessity of laying down these high norms for the Satyagrahi. Satyagraha is not to be resorted to without necessary preparation. Gandhi is quite aware of the risks and dangers involved in using Satyagraha as a mass movement. If each and every individual or a group of Satyagrahis does not fulfil his rigorous requirements, he fears, the movement may degenerate into violence and confusion. One's failings may entail others' failure. He called off Chauri Chaura Satyagraha simply because some of the Satyagrahis deviated from his well-defined path and resorted to violence, killing some police personnel. By Gandhi's own standard, Champaran Satyagraha had been the most successful one. The point is Gandhi is uncompromising in maintaining the high moral standard of political movement. If the moral core of politics is taken away, he warns, it is bound to degenerate into violent power struggle and whatever is touched by violence is tainted by it. His commitment to non-violence is unconditional, uncompromising and absolutely qualitative. To him what is most important is the quality and not number of Satyagrahis. "Strength of number is the delight of the timid." It is interesting to note his strictures against "sitting *dharna*" resorted to by some students during the days of non-cooperation.[13]

> Some students have revived the ancient form of barbarity in the form of 'sitting *dharna*'. I call it 'barbarity' for it is a crude way of using coercion. It is also cowardly because one who 'sits *dharna*' knows that he is not going to be trampled over. It is difficult to call the practice violence, but it is certainly worse. If we fight our opponent, we at least enable him to return the blow. But when we challenge him to walk over us, we, *knowing* that he will not, place him in a most awkward and humiliating position. I know that the over-zealous students who 'sat *dharna*' never thought of the barbarity of the deed. But one, who is expected to follow the voice of conscience and stand even singlehanded in the face of odds, cannot afford to be thoughtless.

Gandhi has often been criticized on the ground that he set-up an unrealistic norm of behaviour for political workers involved in the mass struggle for national liberation. Gandhi's answer has always been the same. The question of scale or quantity, the question of defeat or victory, are all relatively unimportant to him. Moral issue is non-negotiable. And what he calls God's law, the very basis of Satyagraha, is non-violable. To him Satyagraha is a *Dharmayuddha* (crusade for the vindication of the moral order) and every Satyagrahi must strictly adhere to the law of this holy war. He does not rule out the possibility that a non-violent movement may turn out to be violent at a later stage. He said that it was not possible to drop non-violent non-cooperation because of the danger of its becoming violent.[14] For in that case, he argued, we may have to stop the struggle for freedom simply because of the dangers of its possible abuse. Since Gandhi is not a system-builder, as I said before, all his utterances need not be taken literally and should be understood with specific reference to their historical contexts. For elsewhere he has said that he does not mind national freedom itself being delayed by several centuries rather than achieving it through violence and communal blood bath. Gandhi was never unaware of the inherent difficulties involved in launching Satyagraha with its all strict moral conditions at the political level. Bearing the sceptic's reaction in mind, he has spoken of five stages through which the movement would be required to pass.[15] At first the people will ignore or be *indifferent* to Satyagraha. Then they will *ridicule* it. At the third stage they are likely to *abuse* it. Then attempts will be made to *repress* or suppress it. And only at the last stage, when its implication and strength are fully realized, Satyagraha will be *respected.*

There are different modes and levels of human thought and action.[16] One may think superficially, depending on uninterpreted or intellectually using precise and comprehensive concept, or spiritually or intuitively. Similarly, one may act impulsively without caring to assess the value and possible consequence of his actions, or rationally taking into account ground as well as consequence of his action conscientiously, or intuitively dispensing with rational calculation and depending on an inner power which is credited to have all the merit of reason and avoiding all its demerits. Whether the different modes and levels of thought and action are quite exclusive or interpenetrative have been discussed and debated by the rival schools of thinkers down the centuries. But it seems that there are good reasons, both reflective and pre-reflective, to believe that human thoughts and actions may be differently categorized in terms of their different "springs" or sources, orientation and end-consciousness. Many thinkers, especially mystics like Sri Aurobindo, have spoken of three types of thought and action—instinctive, intellectual, intuitive or suprarational.[17] Further analyses would reveal few more grades in between the three referred to here. Gandhi's Satyagraha is a moving plea for spiritual or suprarational modes of thought and action. While he asks the people to follow their inner voice or the Law of God, over-ruling or violating, if necessary, the law of the land and other conventional norms, he in fact asserts unequivocally the superiority of spiritual action over the other types of action, the motive forces of which are not stable and their end-consciousness not clear. His emphasis on Soul-Force rather than material or physical forces is indicative of his preference for spiritual consciousness and spiritual action.

What makes an action spiritual or suprarational? What is the law of God? Why the Law of God should be given precedence over the law of the land? These are very fundamental questions.[18] For one can always do something wrong under the cover of his "inner voice". Or in the name of God he can violate the law of the land and question conventional norms of social life and everything traditional. Erring conscience is *not* a Chimera. A man may be quite honest, sincere, and yet mistaken. History abounds with examples of people who are committed, inspired or God-intoxicated but even then caused incalculable harm to their own ideal. This line of criticism perhaps could possibly be met by referring to the universal character of the Soul-Force. Though apparently individuated or even fragmented in different human beings, Soul is claimed to be a universal metaphysical principle. Therefore, Satyagraha has been characterized as universalizable.[19] The fact that it is not being practised universally, Gandhi might argue, is not a disproof or disconfirmation of its universal truth-claim. Pressing the point further he might have argued, and in fact he did argue that quarrel, conflict and war between man and man, between group and group, and between nation and nation have resulted basically from their failure to realize their identical and universal origin, the Universal Soul, or God.

This is a perfectly plausible line of argument provided one is prepared to accept the metaphysical foundation of Satyagraha, the doctrine of God and the resulting Soul-Force, in the absence of any supporting empirical evidence. Gandhi is a rationalist but not a dialectician. He is prepared to present his position supported by rational considerations but is not inclined towards "logical logomachy" or theoretical disputes. Gandhi's view is to be gathered more from the meanings of what he says than from the sayings themselves and also from his "experiments with truth" and their practical consequences.[20] To say this is not to minimize Gandhi as a thinker but only to indicate his limitations as a system-builder. Many of the greatest systems of thought and religion have been developed over the "disconnected" ideas uttered, lived and left behind by great souls like the Buddha, Confucius, Christ and Muhammad. To find out the "missing links" in between their ideas one has to look into their own lives rather closely and sympathetically to start with. The great social reformers and religious teachers are not generally found to be acute analysts or systematic thinkers. They are gifted with rare insight enabling them to go into the depth of contemporary ills and evils of society, to ascertain their causes, to find the possible ways out, and which inspire them to initiate fearlessly the necessary course of action first in their own life and then in a small circle of committed people. To theorize over a luminous idea or a set of ideas does not require a man of first-grade genius; it could be left to a group of gifted interpreters and followers. But to strike the luminous idea itself, revolutionary in implication and changing the course of human history, is an act of rare genius.[21] A great scientist is not necessarily a good philosopher of his own science. A good poet often commits elementary mistakes in course of giving theoretical interpretation of his own work. Similarly, Gandhi's own ideas of Satyagraha may not provide the best possible interpretation of the same. Perhaps Gandhi himself was not unaware of this possibility. He himself says that one need not take his words very literally or seriously for he might have uttered or written them under momentary impulses or temporary considerations. His famous saying, "My life is my message" is indicative both of his humility and insightfulness. One's utterance or writing may be mistaken or wrong and if one is judged on that score alone, the judgement is likely to be partial and unfair. But one's whole life cannot be a wrong indication of one's own true identity, character and ability.[22]

Gandhi is to be judged primarily in terms of his actions. It is amazing to note that every detail of his life, private and public, big and small, was carefully planned and steadfastly followed. Gandhi the author of the *Constructive Programme,* is not less significant than Gandhi the Leader of the Nation. He did not engage himself only and exclusively in the larger issues of the national life. He was immensely interested in every aspect of it which negates or impedes the development of the individual. In fact he was both a great political leader and a great social reformer. He was deeply concerned with such social issues as communal unity, untouchability, adult education, village improvement, peasant upliftment, development of non-violent trade-unionism, economic and social equality, and de-centralized economic production and distribution through the promotion of cottage and small scale industries. This shows his keen awareness of the fact that unless social transformation is brought about, political *freedom* by itself will not mean anything very profoundly significant. A careful analysis of his constructive programmes makes it abundantly clear that he intended to use all these programmes as potent instruments of social change. A caste-ridden and highly stratified society cannot respond to the challenge of time.[23] He was against untouchability not only on moral ground, which was of course there, but also on socio-economic grounds. Social mobility without education is an impossibility. Education exposes man to new forces and enriches his personality, enabling him to make use of his knowledge to better his life and environment. Gandhi was not happy with the stereotyped and the alien type of our education. His novel scheme of education, *nai talim,* tries to strike a *balance* between *theoretical demands* and *practical necessities* of life. Education with him is not a mere theoretical pursuit or intellectual adventure. It must enable man to develop all aspects and potentialities of his being.

Man is the main dynamo of social change.[24] Gandhi's main concern is the human agent who will mann and execute the constructive programmes. He insists that man must be a Satyagrahi primarily inspired and motivated by his own Soul-Force. This is individualistic to start with, and moral accent of Gandhi's thought is clear from his conceptions of Non-Cooperation, Civil Disobedience and Quit India. It is also evident from his *Khadi and Sarvodaya* movements. In other words, whether we look at the macro-canvasses or micro-spots of Gandhi's thought and action, his emphasis on individual man of high moral quality is unmistakable. When a man approaches the problem of social change primarily through human agency and not through institutional framework, certain corollaries follow from it. Man being universal in nature cannot achieve anything of abiding importance and value through conflict or war. Anxiety and tension, jealousy and hatred, war and crime, are symptoms of man's isolation from man, of his failure to realize his essential universal nature. True change originates from within. True change is non-violent in character and universal in scope. An individual or a group cannot achieve change for better without taking others along with himself or itself. Spiritual motion in an individual being brings about a spiritual commotion in the community as a whole.

Once the main dynamo of social change is shifted from man to institution, we take the wrong road to *Sarvodaya* literally, all-round development. The good that concerns everybody should be consented to and contributed by every body. It cannot be a gift from someone to the rest. It cannot be perfectly cast into any institutional mould. Primacy of institutional approach towards social change tends to perpetuate social conflict. If group-structures or class-structures become primary object of our interest and action, the individual beings therein are bound to be relegated to a secondary position. The common denominator of social aggregates is inversely related to their numbers. The factors which are common to all social groups and institutional frameworks are bound to be few and superficial. Human beings, on the other hand, however large might be their number, have something basically common to each of them. Sometime it is called Soul or Spirit. Sometime it is called rationality. The point is that common living factor can provide a basis of human unity which cannot be provided by any institutional framework, however carefully it might he planned and constructed. Elements of impersonality are bound to inject a mechanical character in the institutional framework and, therefore, institutions have an inherent conservative nature in them and prove relatively impervious to the new forces and agencies of change. Man can change himself more easily, steadily and continuously without distorting or destroying himself. There are inbuilt organic capacities in man, enabling him to adjust himself to other human beings, institutions and environment. When institutions are exposed to the strong forces of change, they are terribly disturbed, often badly damaged, and in some cases even destroyed.

Gandhi is, therefore, against discontinuous and violent change. His non-violent revolution ensures continuance of tradition, preservation and transvaluation of older values, facilitating the process of development of new values. Non-violent Satyagraha not only seeks to maintain the inner continuity between tradition and modernization but also between the classes and the masses, between the elite and the poor, the *harijan* (literally, God's man). The Universal Spirit is spread both over time and space, animating both social space and historical time.

The main basis of Satyagraha as a peaceful method of social change is, as we have noted before, a metaphysical principle, viz., belief in Universal Spirit or God individuated in different human beings without destroying or even tainting their basic identity and unity. The fundamental critical questions raised in this connection are somewhat like this. What happens if this very belief in the metaphysical principle is disputed and rejected? What then will bind man with man, group with group, class with class, and nation with nation, in face of their difference and conflict? What justification there is left for the hope to universalize Satyagraha, bringing all individuals and aggregates together? All these questions, broadly speaking, may be answered in two different ways : one, in terms of the presence and operation of the Universal Soul in the individual souls, and the other dispensing altogether with this metaphysical principle. Something has already been said about the first view, particularly referring to Gandhi's thought, indicating briefly its limitations. I think that the second view also deserves serious attention and careful consideration. Most of Gandhi's ideas and actions in respect of Satyagraha can, I think, be satisfactorily explained without using the God-in-Man hypothesis. The question of God has remained a favourite subject of philosophical disputes from the very dawn of human civilization and culture. The results of human reason and experience accumulated and assimilated over the centuries have left behind enough of norms and forms to minimize conflicts between man and man, and between group and group, and enabling mankind to discover and establish numerous modes of conflict, cooperation and integration. Some of the noblest teachers of mankind like the Buddha and Marx who worked so much for human unity and progress were atheist. Some of the most well-known and influential followers of Gandhi like Nehru did not share their master's faith in God. Satyagraha is a noble principle. But its metaphysical foundation seems to be questionable and dispensable. This I have argued elsewhere.[25]

REFERENCES

1. *Harijan*, 27 May 1939.
2. Raghavan N. Iyer, *The Moral and Political Thought of Gandhi,* Oxford University Press, Delhi, 1973, pp. 260-61.
3. Henry David Thoreau, *Civil Disobedience,* Revell Paperback, 1964; Leo Tolstoy, *The Kingdom of God is Within You: Or, Christianity Not as a Mystical Teaching but as a New Concept of Life,* New York, Noonday Press, 1961; see also Jacques Maritain, *Scholasticism and Politics* (tr, Mortimer J. Adler), Macmillan, New York, 1940.
4. *Young India,* 6 May 1926.

5. *Ethical Religion,* S. Ganesan, Triplicane, Madras, 1930, p. 45.
6. *Young India,* 9 June 1920; see also H.L.A. Hart, *Law, Liberty and Morality,* Oxford University Paperbacks, 1971.
7. *Young India,* 10 November 1921; see also, "The legitimacy of violence as a political act," in Alexander Klein (ed.), *Dissent, Power and Confrontation,* MeGraw Hill Book Co., New York, 1971.
8. *Young India,* 31 December 1931.
9. D.P. Chattopadhyaya, "The paradox of violence," in *Societies and Cultures,* Bharatiya Vidya Bhavan, Bombay, 1973.
10. *Young India,* 9 June 1920; see also P.H. Partridge, *Consent and Consensus,* Macmillan, London, 1971.
11. *Harijan,* 22 October 1938.
12. *Amrita Bazar Patrika,* 3 August 1934.
13. *Young India,* 2 February. 1921.
14. *Young India,* 15 December 1920.
15. *Young India,* March 1921.
16. J. Macmurray, *The Self-As-Agent,* Faber Paperback, London, 1969; see also, Stuart Hampshire, *Thought and Action,* Chatto and Windus, London, 1970.
17. D.P. Chattopadhyaya, *History, Society and Polity,* Macmillan, New Delhi, 1976.
18. See, for example, Roscoe Pound, *Jurisprudence,* Vol. 11, Part 3, Ch. 11, West Publishing Co., St. Paul, Minn., 1959; and Julius Stone, *Social Dimensions of Law and Justice,* Ch. 12, N.M. Tripathi Pvt. Ltd., Bombay, 1966.
19. *Young India,* 3 November 1927.
20. *An Autobiography or The Story of My Experiments With Truth;* see particularly "Farewell", Vol. II, Navajivan Publishing House, Shriman Narayan (ed.), *The Selected Works of Mahatma Gandhi,* Ahmedabad, 1969.
21. Michael Polyani, *Personal Knowledge,* Routledge & Kagan Paul, London, 1958; see also N.R. Hanson, *Patterns of Discovery,* Cambridge University Press, Paperback, London, 1972.
22. J.B. Kripalani, *Gandhi: His Life and Thought,* Ch. 38, Publications Division, Government of India, 1970.
23. David G. Mandelbaum, *Society in India,* Vol. II, Ch. 23 and Appendix, University of California Press, London, 1972.
24. D.P. Chattopadhyaya, *Individuals and Societies: A Methodological Inquiry,* Ch. 1, Scientific Book Agency, 2nd edition, Calcutta, 1973.
25. See (17) and (24) above, and also, *Societies and Cultures,* Bhartiya Vidya Bhavan, Bombay, 1973.

12

Sociology, Ideology and the Indian Peasantry:

A Case Study in Social Growth and Stability

I

PRELIMINARIES

In this paper, I propose to discuss the question of growth and stability of India mainly in the socio-economic context. However, at the end of the paper, I would add a few words from the political point of view.

I have decided to focus my attention on the rural life of the peasantry of India primarily for one reason. The real India, as Mahatma Gandhi used to say, is to be found in its villages. Now, 59 years after independence, when urbanization and industrialization have exerted considerable influence on the country as a whole, perhaps Gandhi's statement might sound a little exaggerated. Even if his basic point is accepted, it is difficult to deny that the factors responsible for the stability and growth of the Indian society are clustered clearly around the life of the people who live in villages and whose main livelihood is agriculture and other allied activities. I would discuss mainly the rural life of India without ignoring, however, its relation with the urban life, which is coming up and spreading fast.

Broadly speaking, there are two sets of factors accounting for the growth and stability—sociological and ideological. Human action and motivation are partly influenced by sociological factors and partly by ideological ones. Social customs, conventions, institutions, etc. somewhat manifestly shape human motivation and action. The ideological factors—values, commitments, human needs, etc. are not so manifest in their operations. Analysis reveals that human action is the confluence of sociological and ideological factors. Negatively speaking, human action is not shaped exclusively either by social customs or by social ideals and commitments. It is interesting to note how the two sets of factors interact in shaping the social life. I shall avoid almost dogmatically two extreme positions: (a) ideologization of facts or planning growth without toil and tears, and (b) sociologization of values, i.e. putting an end to ideology and leaving planning to the "hidden hand of nature'. For each one of these two positions has received extensive analysis at the hands of various other writers. Besides, this is not the main objective of my present paper. It would be found that ordinarily I am for *orderly change.* And by "stability" I mean a state of dynamic equilibrium as distinguished from a status-quoist position. The values of the variables of a model of dynamic equilibrium must change in their magnitude according to social needs, particularly those of poorer sections, for a faster rate of growth. Let me make it plain at the very beginning that I do not propose to offer a model of socio-economic growth. Mine is a theoretical analysis which would indicate the socio-economic requirements of growth.

II
THE PEASANTRY OF INDIA: ITS IDENTITY AND TRANSFORMATION

The Indian peasantry is very much a part of the Indian social milieu. India's economy being predominantly agrarian, the problems and prospects of the peasantry will continue to rate high among the factors facilitating or impeding the progress of the country as a whole. In the economic classification of the workers of India (1971), cultivators account for 42.9 per cent, agriculture labourers 25.8 per cent and other workers 31.4 per cent, the total numbers of these three categories being 78.7 million, 74.3 million and 57.6 million respectively. Of the total population of 548 million, 438.6 million live in the rural areas. But the number and percentage of population living on land are going down. Correspondingly, the number and percentage of the population depending on industry and urban-based occupations are increasing. In 1972-73 the work force employed in agriculture and allied activities was 74 per cent and that in manufacturing and allied activities 18.7 per cent, and in 1977-78 the former came down to 70.7 per cent and the latter went up to 21.5 per cent (Sixth Five Year Plan, 1980-85, Annexure, 13.8).

One can look at the peasant society in two different ways, diachronically and synchronically. As a synchronic segment of society, identity of the peasantry has to be gathered primarily as a product of history and tradition. The synchronic study has its own advantages. It enables us to look closely into the static factors which account for the stability and structure of the whole society. Here I am using the terms stability and structure in a purely descriptive sense and not in any evaluative one.

To understand the peasantry properly we must look at the peasant against his social background. Social background is an umbrella term for a very comprehensive concept. It includes so many concepts under it—(a) *family*, (b) *caste*, (c) *community*, (d) *class*, and (e) *system*.[2] The identity of the peasant assumes and loses different traits in relation to these different reference groups. The disposition and the behaviour of the peasant in his family context are somewhat different from the same in relation to his caste or community reference groups. The comprehensiveness of the reference groups and the subjectivity or the objectivity of the psychological factors could be definitely corelated in a graded scale. This sort of study has its own intrinsic merits. Firstly, it could exhibit, for example, how the peasant is opening up to, or withdrawing from, his immediate and remote reference groups. Secondly, it can bring out the motivation and orientation of his individual action and of the concerned group action. Thirdly, the results of this synchronic study can also be fruitfully exploited in the diachronic context, indicating how and to what extent his motivation and action influence the process of social transformation. Finally, these five reference groups, taken as five graded social units, may also serve as excellent explanatory categories, enabling us to tackle different types of sociological problems. For example, family analysis of the peasant might shed considerable light on the labour, i. e. economic, demographic, and other related problems like nature and size of the land holdings. Caste analysis and community analysis may indicate, among other things, voting pattern, political trend, political leadership composition and other factors facilitating and retarding integration of the reference group in relation to the rest. It has often been pointed out that explanation of action and motivation of any individual, in this case of the peasant, in terms of his immediate reference group is bound to be inadequate, if not fallacious. Because the family identity of the peasant or, for that matter even his caste identity, is not the only identity he is subjectively and primarily conscious of; he has his other identities, e.g. religious and linguistic.[3] While for the analytic-explanatory purpose, we might objectively highlight his immediate reference group identities, but underestimating or ignoring altogether his *dominant* and *subjective* identity consciousness, one might pertinently and critically observe here that sociology is after all sociology and must not bother itself so much with the individual psychology or the group psychology.

To obviate the difficulties involved in this type of immediate reference group analysis, it has been suggested that we should resort to more comprehensive analysis, e.g. analysis in terms of class or in terms of social system. Class analysis or system analysis is unexceptionable in principle. But it serves only a limited analytic purpose, though not any explanatory one. Analysis is not necessarily explanatory. It may or may not be. If the peasantry itself is not class-conscious or system-conscious, then a class analysis or system analysis of the peasantry might serve a purely analytic purpose from a particular theoretical point of view. But on this basis, an action programme, even if drawn up, cannot be properly executed.

III
STATIC FACTORS

Now let us look into the *sociological factors* responsible for the relatively static character of the Indian peasantry and the relative lack of its consciousness of the remote reference groups, class, and system. Among the various factors I would like to mention only few: (1) tradition, (2) religion *(dharma),* (3) custom, (4) convention, (5) rite, (6) myth, (7) *Karma* (action) (8) *Samskara* (deep-rooted impression and disposition) (9) rebirth, and (10) Hinduization or Indianization.

Each of these factors merits detailed discussion. But I propose to offer only few analytical and critical remarks. *Tradition* is hangover of history without disclosing its rationale. It is more regulative and prohibitive than creative and prescriptive. A traditional society or community is generally backward-looking and therefore static in character. I say "generally" because there are non-static types of tradition as well, e. g. scientific and technological traditions. While in the former case traditional actions are imitative and repetitive, in the latter case they are innovative and creative. The same is more or less true in respect of religion. Religion is a bad English equivalent of *Dharma. Dharma* has legal theological and ethical implications. Codification of religious ideas and institutionalization of religious activities are primarily meant to preserve dead or dying value system, lending them metaphysical sanctions. *Customs, conventions* and *rites* converge on a common end. They embody certain forms of action and in most cases these forms are observed or followed without questioning their content, meaning or consequence. Whenever, these questions are raised one is referred to the sanctions of tradition, *Dharma, Sruti* or *Smriti.* The questions which are relevant in respect of customs, conventions and rites, are relevant in respect of the sanctions as well. When I say this I, do not intend to deny the positive aspect of *Dharmacara* and *Sastracara* actions enjoined by religion and scriptures.

A society bedevilled by transitional anomaly is required to provide some action parameters, however questionable they might be, to its members. Even a normal society has its do's and dont's and provides some guidance to its people to act and respond appropriately under different circumstances. These are also *mechanisms of adjustment and adaptation.* Directly, they are of little or no use in the matter of bringing about *social transformation.* All these factors objectively favour *status quoism.* The concepts to *Karma* and *Samskara* lend strong support to the *status quoist.* According to the concept of *Karma,* a man, whatever he is or has, is the result of his own action. It is a sort of inexorable moral causation and determinism. Every *Karma* (action) has its momentum (after-effect) and *Samskara.* I know it is quite possible to offer a non-deterministic interpretation of the concepts of *Karma* and *Samskara.* But that will be theoretical rationalization of these concepts and not how in fact they are accepted and practised by the bulk of the people of the country. Re-birth is an extension of the concepts of *Karma* and *Samskara.* The producer of *Karma* must also consume its effect, if possible, in this very life and, if not, then in the life beyond. It has been said that the *Samskaras* generated by *Karmas* are so deeprooted in the human being that even if the gross body *(sthulasarira)* is destroyed, the *Samskaras* remain active and effective even in the subtle body *(Suksma-sarira).* These concepts, as anybody can well see, are too metaphysical or transcendental to be empirically tested. But the layman's life and the commonsense view of life, in spite of their metaphysical origin and inspiration, are not reflective and critical in character. Popular *faiths* and *myths* are enough to persuade the common man to accept the traditional life without question and query. For example, the Indian myth of solar eclipse or lunar eclipse (that a trunkless demon, *Rahu,* devours the sun or the moon temporarily and then the latter comes out of the lower part of its cut-off throat) is enough to persuade him to accept it as valid, and therefore he does not bother the astronomer to tell him what we call the scientific explanation of this "strange" phenomenon. It is thus by uncritically accepting certain authoritative modes of thought and action that the common man easily adjusts himself to his own social milieu. His uncritical eagerness to belong to his social unit and identify himself with it makes him feel easily inclined to *Hinduize* or *Islamize* himself.

This brief analysis of the static factors of the society is more or less true of all groups of people of the Indian society, depending, of course, more or less on the level of their education and the degree of their exposure to the urban-industrial influences. The peasant is interested more in the preservation of traditional values than in the transvaluation of the values handed over to him by the "wiser and older" generations and accepted by him in good faith.

IV
DYNAMIC FACTORS

The other side of the picture of rural India and of peasantry has so many *dynamic factors* in it and is extremely complex.[4] Like the static factors the dynamic ones are also of two types: conceptual and institutional. Of the institutional factors most important perhaps are : (1) industrialization, and (2) education. Our educational system has been fashioned out and out in the British model. Since the impact of this education is not intensively felt beyond the urban and semi-urban areas, the peasantry has not been deeply influenced by it, except of course through its inevitable spill-over effects. The more the traditional systems of education have been replaced by the modern and completely alien model of education, the more the traditional system of life has been. It is true that education has worked also as a *modernizing* factor. The alien character of the modern education is easily perceptible from its after-effect. If a son of a rich farmer is educated, say, for 5-6 years in an urban environment and becomes a doctor or engineer, he hates to go back to his original or any other rural place. This is even more tragically true in respect of the students educated abroad.

Our education because of its alien character instead of making us conscious of our "true" identity, facilitates the process of alienation inherent in our feudal and capitalist modes of production. Given its present form, it will take away the student from his traditional mooring and objectively promote alienation. It is perversely related to the process of social change via alienation.[5] Education has become a machine for printing job vouchers. Most of the job opportunities opened up by this largely irrelevant education system have hardly anything perceptible to do with the national perspective of development. Generally speaking, its relation with different vocations is very weak or narrow; in most cases both. The impact of this system on the peasantry is to a great extent negative. When I say all these, I do not propose to deny that even the alien system of education imparts a sort of skill or understanding or both which may be positively related with the trend of the increased agricultural production. But that does not contradict my basic contention.

When the products of this education system run institutions like schools and banks in the rural areas and administer projects like (3) the Community Development, and (4) the Rural Health Centres, including the Family Planning Centres and Camps, the outcome is bound to be mixed, partly positive and partly negative. These projects and concerned personnel inject undoubtedly an element of dynamism into our relatively static peasant society. (5) The extensive network of roadways and communications, and (6) the increasing coverage of the mass media are also partly responsible for breaking the cultural inertia of the rural society. In this connection we should better bear in mind that the physical and mechanical facilities of communication by themselves are not sufficient to remove the cultural gap between the urban society and the rural society and ensure uninterrupted and effective communication, facilitating the process of cultural integration and undoing, at least partly, miseffects of stratification. (7) Generally speaking, the increasing sphere of State activities necessitating deployment of increasing number of officers and technologists and their complementary contingent is also strengthening the forces of social change in the rural area. If the resulting change is not that perceptible as one would have otherwise expected of the mighty State initiative, the causes are not far to seek, viz., the negative, insensitive, and sluggish characteristics of administration, unimaginative plans and programmes, lack of popular participation, and extremely poor level of peasant organization and movement.[6] In other words, those who are officially responsible for introducing change in the rural life are not subjectively interested in promoting, organizing and strengthening the rural leadership. The average official's interest in the rural upliftment is typically oriented by his alien education and petty-bourgeois bias.

Among the conceptual factors of social change I should mention (8) Renunciation (*Tyaga* or *Vairagya),* (9) *Yajna* (act of offering oblation, or sacrifice) and (10) *Moksa* (liberation or salvation). The motivation underlying these concepts is self-exceeding, altruistic, and, in a way, opposed to vested interests. These concepts, rightly understood and practised, promote social *integration* and mobility. But, as it happened in the other cases of the concepts of *Dharmasastra,* the feudal mode of production and its accompanying servile morality murdered the original motive-force of these dynamic concepts and in the course of time reduced them into routine rituals.

Substantial incursion of the urban-educated elements and emergence of a thin layer of educated people in the countryside have created a "disturbed" situation there. The composition of the rural people may be classified under three heads: (a) occupational, (b) motivational, and (c) proprietary. These groups are not topologically isolable and do overlap each other at many places. The landless wage-earner is dispositionally a pro-changer. For he has nothing to lose but his under-paid servile occupation. In an ordinary situation, not disturbed by the extraneous forces, the proprietary class is motivationally anti-changer. Understandably he is a *status-quoist.* He is in favour of preserving the stratified character of the rural society and of the agrarian economy. According to him, *stratification* promotes *cohesiveness* and *orderliness* in the society. But the picture of life to the poor peasant or to the landless peasant is quite different. He prefers *friction* to deceptive cohesiveness and illusive *integration* of the static society. Friction and conflict generate social commotion and promote the process of socialization, breaking or, at least, weakening the barriers of stratified society. But this process of socialization is resisted not only by the landlord and the rich farmer but also by the professional and occupational groups consisting of village priests, *pundits, Maulanas*, moneylenders and quacks; all of them have their own vested interests in the static society.[7] It is true that each of these groups performs, in a limited sense, a positive role in the absence of prochange professional group. For example, desperately in need of credit and with no nationalized bank near about to come to his help, the poor peasant is forced to go to the village money-lender, mortgage his land and hypothecate his standing crop on the field for a very low consideration. Similarly, the villager with no qualified doctor or hospital near him is obliged to depend on the quack. He has little or no option before him : either he can see his children or wife dying untreated or treated by a quack. He just takes a chance.

Though the people having vested interests in the agrarian economy do not willingly produce and promote the forces of change, forces are *objectively* generated by a mixture of causes. I have already referred to *the forces of transformation and mobility* and spill-over effects of the urban life in the rural area. Besides, the purely urban streaks and shocks reaching the rural life, there operate also some semi-urbanized *feedback peasant motivations.* For example, some landless peasants of Bihar and Uttar Pradesh work for some months every year in the urban and semi-urban jute belt of Metropolitan Calcutta and then return to their village homes with new ideas and motivations. This annual migration of the rural masses and their coexistence with the unionized working class in the urban industrial environment and their counter-migration introducc many prochange factors in the rural life.

Endemic forces of transformation are also there in the rural life. For example, the farmer who adopts *modern methods of cultivation,* using chemical fertilizer and tractors, is primarily interested in increasing his income. But his income and affuence entail something not intended by him. The use of *fertilizer, tractor, pump, diesel, electricity,* and the like exposes him and also those who work under him to different sets of contrary forces. The rich farmer knows what makes him richer, and the poor worker is aware of the causes of his own gradual impoverishment. So under the existing constraints of agricultural economy like *unimplemented land laws,* and *non-existent,* or *unenforced minimum agricultural wages act,* green revolution or ambitious programme to augment the agricultural production cannot yield the desired results. I admit that merely by introducing modern methods of cultivation and without disturbing the static socio-legal framework of the peasant life, we may increase the agricultural production up to a point. However, this method of increase in production works only up to a point and not beyond that. Where and how the point is reached is an empirical question and cannot be generalized *a priori.* We must not be over-impressed by the green patches of the Green Revolution. For the country by and large continues to remain grey in terms of the rate of growth in agricultural production, distribution of the increased income from agriculture and allied activities, spread of education, implementation of land laws, and population explosion. Our population growth rate has outspaced the food production growth rate. What is further interesting to note is that the growing affluence of the rich peasantry has not been proportionately reflected in the mobilization of investable resources as evident from the available documents. The impact of the rich peasantry and its institutional ramifications in the rural areas, politically speaking, is no less important than that of the industrial bourgeoisie on the national scene.

V
ON THE SOCIOLOGY AND THE IDEOLOGY OF THE INDIAN PEASANTRY

Let us now briefly recapitulate the results of our preceding analysis. *Any attempt to draw a sharp line of demarcation between sociology and ideology is bound to end up with blindness of the former and the emptiness of the latter.*

Ideology is the value-orientation of thought and action. One's thought and action are largely shaped by one's own interest. Man is affiliated in a graded fashion to different groups, primary and secondary, immediate and remote. He is *partially* free to choose his ideology in terms of his dominant interest-group or/and reference-group. In a static society his action and ideology are mainly shaped by his immediate reference groups like family and caste. *Per contra*, when due to the introduction and operation of the dynamic factors, conflict and mobility are generated in an inert society, remote reference-groups like class and nation assume live character and start influencing motivation and action patterns of the people concerned.

The rural India is waking up from its age-old slumber. In some parts of the country, one might say, the peasantry has already woken up. Its causes are mixed. So also are its effects. Some of the causes and effects have already been indicated before. The point now is that these causes and effects have to be *regulated in a rational manner* and cannot be left to the forces of "spontaneity". Unless this responsibility of regulation is given to (in fact it has to be taken up by) the people of the concerned group, who are themselves vitally and economically interested in the matter, it cannot be satisfactorily discharged. The interests of the Indian peasantry, cannot be defended and promoted by the urban-based upper-middle class bureaucracy and the political leadership whose education and orientation are alien to the reality of the situation and whose interest are objectively related to the industrial bourgeoisie. Some individual officers or party leaders may be honest and well-meaning but that has little impact on the sad institutional state of affairs.

The operative part of the ideology consists of several factors. First, the structure of our education pyramid should be inverted, extensively broadening the base of primary education and making our education more concretely relevant to our social and economic conditions. Increased accent on vocational, informal, and non-formal education is absolutely called for. Equally called for is a radical change in the existing narrow official definitions of "vocational", "informal", and "non-formal". Secondly, we urgently need induction of peasant and working-class elements in the political leadership composition, enabling it to be more appreciative of and responsive to the genuine needs of the weaker sections of the society and little less pervious to the ideas of the well-organized, city-based, middle-class professional groups.

Besides the said two ideological growth inputs, I think, we also need *retention of integrative power of culture,* a point which has been emphasized by such sociologists as Ralf Linton and Margaret Mead. The processes of modernization, urbanization, and industrialization are bound to unleash certain disintegrative forces. This is the price which, perhaps, every developing society is obliged to pay. The limited question is how that price can be pegged down to the minimum and paid by the concerned people without seriously and adversely affecting their capacity to grow. Already the model of "successful economic man" has captured the imagination of the new generations. Those who are receiving higher education, living in town or exposed to urban influence are found to be out of step with the needs of traditional molecular family. The cohesiveness of family life has been breaking up. The atomic family life is also coming under 'increasing stress and strain.' True, this is not peculiar to the developing Indian society. But it has created special problems for India because of its long and strong cultural mooring. Interestingly enough, the cultural moorings which have given us an added assimilative power, enabling us to assimilate new and different ideological forces of foreign origin, have at the same time generated deep convulsions and confusion. Those countries which are less firmly rooted in their past, experience less cultural confusion and social convulsion in the process of economic growth.

For accelerated economic growth with social justice, additional inputs are necessary. We need more investment from internal savings. The Sixth Five Year Plan envisages a total outlay in the public sector of Rs. 97,500 crores at the 1979-80 prices. Discounting the rise in price level, this outlay represents an increase of more than 80 per cent in comparison with the Fifth Five Year Plan outlay.

But this does not seem to be enough to alleviate the poverty of the poorest sections of the people in the rural sector. The alleviation of rural poverty, it has been said, will be the prime objective of the current (Sixth) Plan. For ensuring effective solution of rural poverty what is primarily needed is an increase in the productive potential of the rural economy. Because of the inadequacy of resources for higher growth rate in medium-term, more direct means of tackling the problem of poverty and destitution have to be resorted to. The main strategy for effectively tackling the medium-term problems are : (i) increasing production and productivity in agricultural and allied sectors; (ii) training programmes to promote self-employment and wage employment amongst the rural poor; (iii) making available the credit necessary for supporting the programmes taken up for the rural poor and effectively regulating the exploitative role of the village money-lender; (iv) promoting the marketing support and *gradually* regulating the role of intermediaries; (v) involving educational and industrial institutions in preparing a shelf of projects for gainful self-employment and for drawing up plans for scientific and economic utilization of local resources. The official planners themselves have recognized most of these requirements. But the requirement which seems to me very basic is yet to receive adequate recognition; and that is the requirement of *equitable distribution of the income from increasing agricultural and allied production.*

Of the total domestic saving of Rs. 149,647 crores, public saving accounts for 22.9 per cent, and the balance of 77.1 per cent represents savings generated in the private sector. The net inflow of external resources to the public sector plan has been estimated to be Rs. 9929 crores and constitutes about 10.2 per cent of the total public sector plan outlay. The net aid is likely to be Rs. 5889 crores.

Additional resources have to be mobilized by (a) appropriately taxing the agricultural sector; (b) reducing subsidies (in respect of use of electricity and fertilizers, for example); (c) ensuring profitability or at least reducing the losses of Public Sector Corporations, State Electricity Boards, State Transport Corporations, etc. Some hard decisions have to be taken for the maintenance of firm fiscal discipline.

The estimates of financial resources totalling Rs.92,500 crores leave a gap of Rs.5000 crores for financing public plan outlay. Can we make it up by deficit financing? Perhaps, we can. But, this gap is likely to increase. Besides *manageable* deficit financing, we should also think of getting chore foreign aid. Even the communist countries now opt for different forms of foreign aid. Compared to many other developing countries, India has a very good case for receiving more foreign aid. The per capita foreign aid received by India is extremely poor. The India's case for more foreign aid has been persuasively argued by many distinguished economists like Myrdal, Viner, and E.A.G. Robinson. Unfortunately, our case has not been duly appreciated by the governments of some rich industrialized countries.

It seems to me that a very important institutional input necessary for the orderly growth of India is a *responsive* and *accountable* bureaucracy. Because of the weakness of the political party system and consequential instability of the State governments, which, one feels, would remain for some time to come, bureaucracy is called upon to play a very important role. This role may be played both ways, constructively with a growth orientation and negatively in a routinized and sluggish manner. Because of the nature of education and training of the members of our bureaucracy, generally speaking, they are not adequately responsive to the needs of the poorer sections of the people. By imparting a rural orientation to their training, making them more accountable, and freeing them from illegitimate political pressures, I think, this problem can be solved at least to a certain extent.

We must encourage *innovative people* all along the line, in every sphere of national life—administration, industry, agriculture, education and technology. Our main resource is manpower. And we must try every possible means to improve its quality in different directions. One of the desiderate of improving its quality is to control its size. Its size was 609 million in 1976. It is estimated to be 735 million in 1986. At present the birth rate (per thousand population) is 33. To reduce the net reproduction rate to unity by 1995, the birth rate has to be reduced to 21, death rate to 9, and increase the proportion of couples protected by family welfare planning to about 60 per cent. And this is not an easy task at all. A policy of carrot-stick-mix has to be imaginatively formulated and implemented with dedication and vigour.

VI
CERTAIN GEOPOLITICAL FACTORS AND THE PARADOXES OF THE SITUATION

In the shrinking world of today, it is almost as a matter of course that the growth and stability of one country is shaped, to a considerable extent, by what is done and not done by other countries, i.e. by the forces of foreign origin. In the case of India this general rule has proved to be true in a very special manner. If we look back into the cultural past of India and the stage of its dominant economic mode of production in the first half of this century, one feels inclined to agree with Mahatma Gandhi that the independent India would not be well-advised to ape the democratic system of the White-Hall-and-the-White-House type. Those who were critical of accepting the political system of the western origin or sceptical about its possible outcome were not necessarily critical of the underlying ideology as such. Their main point was that, given the Indian situation, this system might not prove the best possible one. It must be admitted here that it is very difficult to indicate clearly the criteria for rational choice of political system. The choice of the people of a country as a whole is very unlikely to be unanimous or homogeneous. The choice of the rich and that of the poor are not likely to coincide. Even when they say in the "same" voice, "we want freedom" or "we want democracy", they do not mean the same thing. Primarily because of the ideological orientation of its political elite, India, like many other developing countries which earned their independence after the Second World War, opted for the White House-and-White-Hall model of democracy as defined by its five parameters: (i) multi-party system; (ii) independent judiciary; (iii) periodic elections; (iv) freedom of trade unionism; and (v) freedom of press. The leadership of most of the developing countries responsible for adopting western democratic system was mature, patriotic, honest, and had gone through the hard struggle for national independence. And yet, interestingly enough, historical experience of thirty years or so has made it abundantly clear that the political choice of most of the newly independent developing countries was wrong or unworkable. Individual case studies would show that honest political decisions of the political elite was not in consonance with the *deeper needs* of the people of the concerned countries. Consequently one finds today that most of the countries which had originally opted for the western model of democracy could hardly satisfy the said parameters of the model. Most of the countries have given up the multi-party system. In many cases the freedom of press and/or trade unionism is in effect being denied. The independence of judiciary, though formally recognized, is a widely suspect notion. Similarly, periodic elections, even if held, do not enjoy the credibility of the people concerned. While I mention these limitations *in practice of* the western model of democracy in the context of developing countries, I do not propose to criticize the model as such or question the intentions of the leaders who opted for it. Men may be, often are, honest and yet mistaken. The main lesson that we gather from the historical experience of the last thirty years or so is that the political ideology or system of a country should not be decided in an abstract manner, i.e. disregarding the *deeper needs* of the larger and the weaker sections of its people.

Compared to many other developing countries, India satisfies better the parameters of the western model of democracy. I say this neither to flatter ourselves nor to criticize other developing countries. Far from that, objectively speaking, the situation appears to be this. As India embarked upon the process of industrialization and introduced the western system of education in the middle of the nineteenth century, it was at a relatively advantageous position in terms of administration and technological manpower to maintain at least the forms of western democracy. But we ourselves know where lies the weakness of our system.

Our geopolitical position has a deep impact on our political system and its underlying ideology. Two of our neighbours, China and the Russia, have opted for slightly different forms of the Marxist model of democracy One of them is an acknowledged superpower and another, one might say, is in the process of being a superpower. Ours being a democratic system of the western type, is *open* even to its avowed opponents. But a democracy of the Marxist type, the Russian, for example, as they themselves say, is not open to its opponents. The influence of the Marxist ideology, though slowly, is spreading in the country especially among the educated urban youth and, of late, the rural poor and the industrial worker. The Marxist ideology has aroused a high expectation for faster growth. Many of us, including the Marxists committed to the ideals of the Dictatorship of the Proletariat, are quite aware that in the context of our relatively backward economy, it is difficult, almost impossible, to ensure fast economic growth without compromising, more or less, the said characteristics of the western model of democracy. Briefly speaking, the paradox of the situation is this. If we want fast economic growth some constraints have to be imposed on the people; and if all (western) democratic rights are granted, the growth rate is bound to be slower and the war against poverty cannot be won quickly. In our national context it is of no use to remind us and point out, as some people do, that there are instances of fast economic growth coupled with all *forms* of democratic freedom. The cases of such countries as Japan and, the USA, on analysis, are found to be not of much relevance to the context of our country. Those countries, unlike ours, had an opportunity to develop their economy *over a very long period* unaffected by foreign rule and serious exploitation. Through education and industrialization, they could develop a strong manpower base. Besides, *during the most crucial period of their economic development,* they did not face the forceful Marxist challenge, either geopolitically or ideologically.

The case of India is in a sense unique. Committed to the western model of democracy, burdened with a very big population, and, given the static factors of its society, India finds it extremely difficult to achieve fast economic growth to the satisfaction of all concerned, particularly to that of the poorest sections. And to the latter Marxism for understandable reason appears to be an attractive ideology of freedom from poverty. The relative merits of the alternative ideologies and the growth strategies would probably be decided in the medium-term primarily by *historical experience* (as interpreted and understood) and not by *abstract arguments.* I know this ("historical experience *vs* abstract arguments") formulation is not happy; but, it seems to me, this is how the people, in general, and the poor, in particular, are going to take the matter.

REFERENCES

1. A.M. Khusro (ed.), *Readings in Agricultural Development,* Allied Publishers, New Delhi, 1968. See, also, J.W. Mellor, *et al.* (eds.), *Developing Rural India: Plan and Practice,* Cornell Univ. Press, Ithaca 968; Mandelbaum, *Society in India: Change and Continuity,* Vol. II, London, 1972; Unnithan, Indra Deva and Yogendra Singh (eds.), *Sociology of Culture in India: Essays in Honour of Professor D.P. Mukherji,* Prentice Hall India, New Delhi, 1965.
2. D.P. Chattopadhyaya, 'Sri Aurobindo on smaller human aggregates: family, clan, commune and tribe," *Sri Aurobindo Mandir Annual,* August, 1973, pp. 46-73; Talcott Parsons, *The Social System,* Free Press, Glencoe, 1964; Kingsley Dais, *Human Society,* Part III, Macmillan, London, 1969.
3. Max Weber, *The Sociology of Religion,* Beacon Press, Boston, 1963. Aaron V. Cicourel, *Cognitive Sociology: Language and Meaning in Social Interaction,* Penguin, 1973; E. Gellner, *Cause and Meaning in the Social Sciences,* RKP, London, 1973.
4. Myron Weiner, *The Politics of Scarcity,* Chicago Univ. Press, 1962; *Modernization: The Dynamics of Growth,* Basic Books, New York, 1966; see also, S . N. Eisenstadt, *Tradition, Change and Modernity,* Ch 12, Wiley, New York, 1973.
5. D.P. Chattopadhyaya, *Societies and Cultures,* Bharatiya Vidya Bhavan, Bombay, 1973; B. Ollman, *Alienation: Marx's Conception of Man in Capitalist Society,* Cambridge University Press, 1971.
6. M.K. Gandhi, *Hind Swaraj;* and *Constructive Programmes, Collected Works,* Vol IV, Navajivan Publishing House, Ahmedabad, 1968.
7. Ernest Mandel, *Marxist Economic Theory,* Ch. IX, Rupa & Co., Calcutta, 1971; Theodore W. Schutz, *Transforming Traditional Agriculture,* Yale University Press, 1964; Andre Beteille, *Inequality and Social Change,* Oxford University Press, Deihi, 1972.

Part IV

Dynamics of Values

Dynamics of Values

In two essays of Part IV, I would be developing the same lines of criticism against *static structuralism* and the same lines of arguments for *dialectical dynamism* as I have been developing in the earlier parts of the book. But the focus of the discussion is new: the nature of values—moral, legal and political.

The classical rationalist theory of values is associated, rightly so, I think, with the name of Plato and claimed to be realist. Its main thesis is: Values (with capital V) are universal, eternal and exist independently of their being known or acted upon by human beings belonging to different cultures. Values of human knowledge and action are contingent upon their "affiliation" to those immutable and objective values. Human knowledge and action, according to this received realist-rationalist view, cannot be logically credited with the power to generate or degenerate Values. We, human beings, are copyists of Values—all sorts of Values—aesthetic, ethical, moral and legal. Our copies or approximations may be good, bad or indifferent. And that is all. If man, either in his individual or social capacity, is admitted to be creator or destroyer of Values, the Platonist argues, we are landed in a relativistic position, giving rise to anarchy and confusion in the field of Values. Obviously the actual state of affairs, in the field, as recorded by historians of different value-disciplines, do not warrant this despair. True, on moral, legal and political issues persons of different persuasions and of different levels of understanding, professionals and laymen, for example, have been debating or even fighting in classrooms, courtrooms and open fields, for centuries. But it is also true, perhaps more persuasively so, that they have been agreeing and settling these issues and cooperating between themselves on many other issues. In brief, one might say, human history is marked both by conflict and cooperation. Moral pessimism is uncalled for. Political anarchy is not a frequent visitor to the arena of history. The elements of goodwill and sanity that we see in an otherwise very difficult world around us, thank God, are to a great extent *human* in their origin. We have had among us a few well-meaning utopians as well. Without denying all these straightaway, a modern defender of Plato might respond by saying that this limited achievement of human beings is due to their awareness of those universal and eternal values which are there in them and the resulting actions. It is agreed that the Platonic thesis may be presented more presuasively in modern idioms than it has been above (mainly for the sake of brevity).

But the substantive difference between the Platonic view and ours remains. Values are basically human, cultural products, and subject to the extremely complex laws of cultural evolution. Man is not a copyist. He is a creator. He does not create out of nothing. Endowed biologically to decipher, store, sort and relate the natural and cultural messages, man can *initiate* action and thought processes. In other words, he creates values both as a *part* of nature *and a partaker* in a cultural process. This and other related points have been persuasively argued, among others, by philosophers like Popper, biologists like Dobzhansky and linguists like Chomsky and Lenneberg.

Values answer human needs and are context-sensitive. The distinction drawn by Keynes between *needs* and *wants* should not be exaggerated: for it rests on an inadequate appreciation of the graduated relation between nature and culture and the systematically ambiguous (or many-sided, graduated) nature of man himself. It may be pertinently observed: what make values possible or what enables man to create (and also destroy) values are already *objectively* there, made available to him by nature (through biological competence) and culture (through language, forms of ideas and norms of action). If this admission amounts to a mild dose of concession to the realist-rationalist view on value, I concede ungrudgingly. But, then, I reiterate, man stands in a dialectical relation to values: he is, simultaneously, a producer and a consumer of values, a creator and creature of values.

In essay 13, "Human Rights, Justice and Social Context," I have tried to point out the shortcomings of the universalist theories of values with reference to the human rights. It is a very old view that human nature, being essentially the same everywhere, is morally entitled to some basic human rights for its own development and, through it or otherwise, also for the development of the society in which it is situated. As a part of it, or independently, it has been argued that these human rights should be recognized and enforced by the law-enforcing authorities of every civil society. Some of the defenders of this view are not content with maintaining merely that these rights are "natural" or "morally universal". What, in addition, they claim is that these rights must be uniformly enforced by the lawful authority. While I fully share the laudable concern of the defenders of the *basic* human rights, I find it very difficult to define exactly what are the *basic* human rights, their scope and, particularly, enforcement conditions. This difficulty is indeed very genuine and not of the sort often used to deny human rights. Most of us, I think, would agree *in principle* that there are certain basic human rights and that every civil society must recognize and, *wherever possible,* enforce the same. The main difference arises on the question of *practice* or of determining the enforcement conditions. The right to work, for example, means one thing in a highly industrialized country like the USA or Japan, where the dominant mode of production is capital-intensive and the population-size, in relation to the potential gainful employment scope in the country, is not very big, and a quite different thing in a developing country like India or Bangladesh, where the population size is very big, the level of industrialization low, and large sections of the "officially" employed people are, in fact, underemployed. The health and housing rights recognized and enforced by the USA are simply unpracticable in most of the countries of the developing world. The fundamental point I have tried to argue in essay 13 is that the principles of the basic human rights should not be formulated in an abstract manner, disregarding the conditions, the social context, of the concerned human beings. Similarly, to my mind, any answer, in *vacuo,* to the question, "what is justice?," is equally unrewarding or useless. Not only for the *administration* of just but also for the *formulation* of *justiciable* principles, I think, we must take attending circumstances into account.

In a sense the theory I have proposed is relativistic and, I agree, it has its weakness. In the name of practical difficulties or attending circumstances if the principles of human rights are compromised, one might pertinently ask, would we be left with any human right in the end? Yes, and I should say why. But, before that, it may be mentioned straightway that denial of the relevance of attending circumstances would deprive us of the best available means of testing the principles of human rights.

Human rights, though relative in the sense of context-sensitivity, must not be construed in the positivist way. For they transcend, to a considerable extent, the context in which they are claimed to be legitimate. Enforceability by the state authorities is not their hallmark. Even if they are not enforceable *now*, they do not cease to be rights or lose their legitimacy. Because their main sanction is moral, they reflect our moral aspirations which are not coterminus with the obtained socio-political conditions. Human ideals, like needs grow and are never completely determined by the ascertainable attending conditions.

This main argument of this paper has been extended to the concluding essay 14, "Law, Morality and Polity". I have been trying, as already indicated, to avoid the extremes of positivism and transcendentalism in social thought. In order to achieve the objective I have traced some anthropological, evolutionary and ecological roots of social thought and the values which are there in it. Man is, as Sri Aurobindo says, perpetually self-exceeding, evolving and interacting, both naturally and culturally. By emphasizing the moral basis of law I want to show, among other things, the narrowness of the positivist view of law, defined mainly, if not exclusively, in terms of state sanction and enforceability, and the collective or social character of responsibility. For doing right things and trying to get the good things, I am of the view, we are responsible not only severally or individually but also collectively. By implication I oppose the view that morality is a personal affair and that it is only to the enforceable laws that we owe collective responsibility.

I have also criticized some intuitionist views of morality. I have Cambridge Platonists, in particular, in mind. Their views are transcendental in the bad infallible sense. Values are questionable, improvable, and, if not humanly sustained, likely to degenerate.

The *eternal* image of polity is inconsistent with our basic principles. Plato, in search of an ideal polity, *idealized* an historical one and then argued at length in self-justification. The theory of structural isomorphism between the soul of man and ideal polity is artificial and untenable. Invoking the theory of natural division in the functions of the soul, a static hierarchy, a reinforced system of inequity (not merely of inequality), is sought to be defended in the ideal polity. Because of strong adverse criticism, at times, the theories of *eternal polity* have been given up and replaced by those of *stable polity*. The latter, one might say, are weaker versions of the former. Instead of the *eternal image* of polity the image that is now being propounded is one of *stability*. Eternity gives way to stability. But their common orientation remains and that is anti-change.

In this connection I have tried to show the close relation between Hegel's theory of state and Kant's, a point which is often ignored for, I presume, the common association of Kant's name with liberalism and that of Hegel's with statism. In spite of his evident liberal leanings one feels sorry to find in Kant's thought a strong defense of the unquestionable authority of the state and, on analysis, it is found to be a corollary of his exceptionless (i.e. universal) Moral Law.

13

Human Rights, Justice and Social Context

In this chapter I propose to show that (natural) human rights and justice have to be understood in their proper social context and that any attempt to decontextualize and thus *universalize* them is bound to prove more or less meaningless. However, this is not to deny their extra-contextual relevance and *general* character.

Can one have any right in the state of nature? If the answer to the question is "yes," we must then admit that natural rights are inalienable and to that extent universal. The idea of natural rights is inseparable from the concept of the "state of nature," and the latter is by no means dead today. Of the historically influential ideas of natural rights, Hobbes' and Locke's are perhaps the best known. The relevance of their ideas today is said to be primarily negative, thus suggesting what to avoid in formulating a modern and desirable concept of human rights; but I think there are positive aspects in their views that merit attention and review.

The natural condition of man is a state of every man being against every man. And this in turn results from what Hobbes calls appetitive, passionate, and self-preserving traits of the human nature. Man is a passionate creature, only secondarily a rational creature.[1] Given a situation in which the available resources are limited in relation to the self-preserving desires of such creatures, a state of perpetual conflict is bound to follow. Hobbes' main doctrine is that it is only through an organized political authority, setting up an artificial condition under government, that this unbearable state of affairs can be ended.

"The right of nature," according to Hobbes, "is the liberty each man hath to use his own power as he will himself for the preservation of his own nature."[2] This liberty is the only right one can have in the absence of a political authority to govern. But its worth is not of much consequence, since in the state of nature there is no authority to defend that liberty or, in other words, to enforce that natural right. Unless "the natural right of every man to every thing" is defined and regulated by enforceable law, nobody is under any obligation to respect others' natural rights. It is the surrender of natural rights to some supreme authority that explains the origin of political obligation. Every man must act or refrain from acting as if he were a party to an agreement to obey that supreme authority unconditionally. However, this authority is limited in one respect: it cannot order a man to take his own life or others to kill him.

Hobbes draws a careful distinction between a right of nature and a law of nature.[3] A law of nature (*lex naturalis*) is a general rule ascertained through reason, by which a man is forbidden to do what is destructive of his life or takes away the means necessary for self-preservation and to omit that by which he thinks his life may be best preserved. His right (*jus naturalis*) consists in liberty to do or to forbear something necessary for the preservation of his own life. The question has been raised whether Hobbes' concept of the law of nature and the right of nature are consistent. If, to quote him, "right consisteth in liberty to do or to forbear whereas law determineth and bindeth to one of them", one wonders how there can be both a law and a right to the same thing. An element of discretion is associated with the concept of right, but law leaves no room for that. Obligation knows no exception, but the same is not true in exercising one's right. On this interpretation, right and law are not only distinguishable but also mutually exclusive.

Even while I have the right to preserve my life, it is true, I may not exercise it. But since natural law, a rule of reason, enjoins me that I should use my power to preserve my life, my decision, an exercise of my discretion, not to use it only betrays my irrationality. Whether I betray irrationality or not, the matter of exercising my right is of secondary importance to Hobbes. What he is primarily interested in showing is that in his model state of nature we do have natural rights. Then by the concept of natural law, a rule of reason, he goes on to show that we agree to transfer our rights to the sovereign political authority and thus to put the state of perpetual warfare to an end. The question of transference of rights makes no sense unless the existence of such rights is first postulated or established.

Hobbes is often referred to as one of the chief exponents of the doctrine of absolute sovereignty and Locke as that of the constitutional doctrine. The implied contention is that the former is almost exclusively concerned with the unlimited power of the sovereign and that his defense of the individual's rights and liberties is deliberately of a secondary nature, as if it were left to the latter to defend the same. I do not totally deny that there is a point in this contention, but one must not overemphasize it. Rather, what is more important is the necessity of trying to understand their views on the matter against their different and definite historical backgrounds. The connection of Locke's ideas with Hobbes has often been "distorted and exaggerated."[4] What Locke really intended is an effective refutation of Filmer's Tory defense of absolute monarchy. One is advised to recall in this connection Filmer's criticism of Hobbes on the alleged ground that the latter's consistent pleas in support of man's natural rights objectively limits the power of the sovereign and encourages rebellious activities. Being the liberal he is, Locke earnestly tries to prove equality and liberty of all men. In this important respect his difference from Hobbes is unmistakable.

Locke's model of the natural state of society is God-ordained. In God's scheme of things all men are equal and free, or, at any rate, destined to be equally free. If man cannot reach his moral destiny of equal freedom, it is either his own or civil society's fault. The power that could enable him to develop his moral potentiality, reason, fully is given to him from the time of birth. The same power, rightly realized, can ensure good governance of the civil society. The law of nature is an expression of God's will and, therefore, intrinsically consistent with freedom and equality of all men. "Law of nature. . is the law of reason."[5] Reason reveals to men the law that forms the basis of their equality and also informs them that they are not subject to any external compulsion or to others' will. Those who violate the law of nature ensuring equality between and freedom of men are liable to be injured or punished by the victims of the violation. Wild beasts and despotic rulers are in different ways impervious to the influences of the law of reason, violate the law of nature, and thus threaten the security of life and tend to destroy the very "common bond whereby humankind is united into one fellowship and society."

Whereas in Hobbes system natural rights take precedence over the laws of nature, Locke accords primacy to the latter. Man is bound to the universal order of things by the laws of nature that *oblige* him to become effectively a moral being, a free and reasonable being. Hobbes' accent is on self-preserving rights. In developing his concept of the law of nature, Locke points out that it *obliges,* him to preserve not only his own life, limb, liberty, and other possessions but also, and *equally* so, those of others.

Liberty and law-based obligation are inseparable. Liberty is native to human nature. The law of nature has put man under the obligation to live freely, to make use of this freedom, to develop a concrete moral life and at the same time a cooperative social life. This very obligation accounts for his "right of freedom"— "freedom of will and liberty of acting." "*The end of law* is not to abolish or restrain, but to *preserve and enlarge* freedom. Both freedom of will and liberty of action are grounded in reason. Any authority, individual or collective, that is not grounded in reason and is not, therefore, itself capable of understanding and following law does not command or deserve its subjects' obedience, because the relation between the former and the latter is law-governed, decided by reason.[7] Man's freedom finds its best expression in freedom of judgment subject to the law of reason.

Locke's is not a purely intellectual conception of natural rights. Property is an external and a concrete expression of man's liberty It is a right that is capable of being exercised on things, and its existence is prior to any explicit agreement or contract between men. Liberty is also a natural property. Locke uses "liberty" and "freedom" interchangeably. He would prefer the proposition, "Man has liberty," to the proposition, "Man is free." Although he says, "man is born free and rational," what he means is that "man has in him the potentiality of being free and rational," and it depends on the character of natural law and civil laws to which he is subject. Liberty, a natural right, is a sort of property that can be *owned* and, therefore, encroached upon or even taken away, *unjustly.* Liberty qua property, like other natural rights, is given by God to *men in common* for their self-preservation and self-development.[8]

To account for the origin of the community ownership of property, the concept of law of nature is enough; but to preserve and enlarge private ownership, men do need civil society and civil laws. For the defense and undisturbed enjoyment of private property, civil society must be brought into being by "compact" and "consent" and vested with due executive and legislative powers. Offenses against liberty and proprietary rights defined by the laws of civil or political society must be punished according to law. Government comes into existence simultaneously with civil society, but its end (the preservation of the rights and liberties of the people) is somewhat limited in comparison with that of the latter, and in the event of its failure to act in conformity with its end, it is dissolved.

The natural rights as formulated by Hobbes and Locke, with their focus on the *equal right* of all men to be free, constitute the moral foundation of human rights as recognized (in principle) in our time. But, also, if the equality of all men to be free is recognized as a right, justice naturally demands its protection. The concept of justice applies to men, actions, attitudes, and states of affairs. Equality of freedom (of all adult rational human beings) is a state of affairs, with both subjective and objective connotations, and as such is a good, perhaps primary, subject of justice. In order to pass a judgment of justice on a state of affairs what we are called upon to look into are the qualities and relationships of the concerned human beings and the transactions they enter into. Justice may be considered under two heads, *aggregative* and *distributive.*[9] An aggregative principle of justice is concerned with the total amount of goods and services enjoyed by a group, and a distributive one with the share of goods and services that different members of the group can enjoy. The distributive principle may refer *to valued mental states* such as happiness and want-fulfilment or to *material resources* such as income and education. The distributive principle itself may be *egalitarian* ("divide the cake equally among the children') or *differential* ("give the hungriest more than the rest"). The egalitarian principle simply lays down the mode of distribution, whereas the differential principle specifies the ground of difference in distribution: the ground may be *need, desert,* or *right*. The most basic character of justice seems to be embodied in the definition: render each his due. Of course, what one's due is cannot be decided in a general or an abstract manner; to arrive at a just decision one's qualities or lack thereof and socio-economic circumstances have to be taken into account.

This general definition of justice seems adequate to take care of the well-known contrast between *legal justice* and *social justice* and that between *private justice* and *social justice.*[10] *Legal justice* in its substantive aspect deals with the punishment of offenses and the compensation of injury by means of enacting and enforcing appropriate laws, and in its procedural aspect states the principles of fair trial, rights of appeal, among others. *Social justice* refers to the distribution of benefits and burdens among the individual members and between the groups of members of society through regulation and by setting up appropriate social institutions—property system, taxation system, public distribution system, restriction of monopoly trade practices, old-age pension, unemployment allowance, and others.

Broadly speaking, there are three different, not necessarily conflicting, interpretations or theories of justice: *right-based, desert-based,* and *need-based.* First, rights may be legal or positive, and moral or ideal. The former is recognized and enforceable, the latter may or may not be recognized and, even if recognized, not enforceable by the lawful authority. This distinction, in effect, is not as clear-cut as it is generally supposed to be in the abstract level of juristic thinking. For the rules of positive law, the accepted basis of legal rights, are not free from practical ambiguity, and often need interpretation before application, and the cases, especially the hard ones, in which the rules are sought to be applied have to be clearly identified. Dworkin's arguments on this point are interesting. Ideal rights often ripen into positive rights through propaganda and public discussion. The claims which appear unjust and are unacceptable as right today may appear just and be statutorily recognized as right tomorrow. Social consensus acting as a natural force hastens the process of ripening of a moral right into a legal one.

Justice is a part of morality. J. S. Mill thinks that the concepts of justice and right are coextensive. "Just claims" are also "right claims"; when they are denied, the right-holder is wronged and therefore deserves some law-sanctioned protection.[11] Of course, just claims in most cases turn out to be weaker than right claims unless they have some independent strong social justification. The time lag between just claims and right claims is indicative of what is called the forward-looking, as distinguished from the conservative, backward looking, character of justice. Because of their moral force or other social compulsion when just claims are accorded legal recognition, the persons related to the right holder in respect of those rights are put under corresponding duties. The right holder, in the sphere of civil law, may or may not press his rights, that is, may demand that the duty be performed or waived. In other words, there is a basic connection between rights and *free choice.*

In addition, rights provide security, and the rightholder rationally expects others to respect his right to something. The main advantages that flow out of positive rights are freedom and security. Obviously, this does not hold in the case of ideal rights that are not as yet legally recognized and enforced. The question of *content* or subject does not ordinarily arise in the case of positive rights, because their *forms* themselves are the object of recognized contract or agreement, and that ensures both secure enjoyment and respect of others concerned. Content is the essence of ideal right, and its moral undertone is unmistakable. Like the rights to life, liberty, and property, natural rights constitute a very important part of ideal rights, with many assuming the dignity of positive rights through the, legal recognition of various types in different societies.

Ideal rights in some societies occupy an intermediate position between positive rights and pure moral rights. For example, "directive principles" of the Constitution of India recognize "the right to work, to education, and to public assistance in certain cases", (Article 41) and explicitly state that these principles "shall not be enforceable by any court ...[but]...are nevertheless fundamental in the governance of the country and it shall be the duty of the state to apply these principles in making laws" (Article 37). Some of these principles (in the nature of prescription) have indeed been followed up by appropriate legislation. But many are yet to get the backing of positive laws. What are still ideal rights in developing countries like India have been recognized in some cases, as positive rights in the developed countries of Europe and the United States. To understand the true content of rights, what is most important to look into is neither the constitution nor individual pieces of legislation of the state but its existing sociological conditions :

how the laws are actually interpreted by the judiciary, followed by the executive, and made practically available to the common man.

The theory of desert-based justice is also beset with numerous practical difficulties. For this is an area in which theory and practice interpenetrate deeply, and the line of demarcation between the two is badly blurred by sociological considerations. It is easy to say "somebody deserves something" or "something is due to somebody," but it is not at all easy to decide how a particular person or group deserves a particular thing (income, reward, or title). Sometimes desert is defined in terms of right (e.g., the man who paid the premiums regularly deserves the insured sum at the end of the maturity period), sometimes in terms of need (e.g., the poor farmers of the drought-affected area deserve to be exempted from the payment of annual land revenue). Taking advantage of a loose definition of desert, one may put forward an unjust claim and demand that it be recognized as one's right. The underlying moral tone (though often very misleading) of the concept of desert has to be recognized first. When we say, for example, "X deserves reward", or "Y deserves punishment," what we leave unsaid (but implied) is the *reason* or *basis* of X's reward and Y's punishment. The reason may be an *action* or *quality of character* of the concerned person. Besides, the question of *social evaluation* of the said action or quality is there. Also, the sociological considerations enter into the matter by the backdoor (in the good sense). Social evaluation and recognition of what one deserves may be prompted by utilitarian considerations, such as "This will induce others to follow his good examples in the services of society,"[12] or contractarian considerations, such as "This is likely to yield a fair distributive outcome" in the non-moral and non-substantive, i.e. procedural, sense of the term.[13]

Doubts are expressed on the question whether *needs* constitute an independent criterion of justice and suggestions here been made that needs are either disguised desert or camouflaged right. If need without the support of right need *simpliciter* cannot be accepted as the criterion of justice, then one has to concede by implication that the needs discussed by social reformers and political revolutionaries, that have not yet been recognized by the concerned authority, are of no significance or consequence at all. Even in a predominantly feudal society not committed to a welfare state, people do speak of the needs of land reforms, of ensuring minimum wages to the agricultural labourer, of rural banks and cheap credit facilities for the poor farmer, of rural health service centres, and the like. Although these needs cannot be brought under the category of rights, they can hardly be dismissed as frivolous and unjust, unless of course the financial and other implications of the needs are quite beyond the capacity of the concerned society. Further, the need for equal distribution of the available resources of a society, such as national income, between the different sections cannot be disputed either, at least not on the ground of justice. At this stage, one may like to intervene and say that the question of equal distribution or even that of equitable distribution cannot be justly settled unless we bring in the concept of desert. Need claims, according to this view, are likely to be made in an arbitrary and self-centred manner, in total or partial disregard of others needs. It is clear that it would land us in a near-anarchic and unjust state of society. To remedy the situation, it has been suggested, need claims have to be backed up or justified by desert. Otherwise, if there are two individuals or groups in a society, one privileged and the other underprivileged, how can we think of distributing something equally between them without sacrificing the principle of justice. In no case should goods be *equally* distributed between the two unless they deserve the same equally. Even for the purpose of equitable, in a sense non-equal, distribution of goods, some rules of desert determination should be followed.

To talk of desert-based needs is not particularly illuminating. For in addition to the difficulties involved in determining one's desert, we have to face here the problems attending the determination of one's needs. Neither look nor words nor social position of a person, certainly not others certificates about him, provide any definite basis for the determination of his needs. Needs may be classified under different headings: *substantive* and *procedural*; *intrinsic* and *extrinsic; original* and *imitative*. They may be illustrated by corresponding need statements in the same order: "Travellers need passports to visit foreign countries", and "The applicant for a passport needs information regarding how and to whom to apply for it"; "Ram needs music and poetry to live his life"; and "Rahim needs colour-matched curtains and cut-glass flower vases for his drawing room"; "We need food, a house, and clothes to live as human beings", and "My children need a colour TV set."

In the case of the need statements of ordinary language, the possibility of cross-classification cannot be entirely ruled out. As Brian Barry[14] has quite rightly pointed out, every need statement of the form "A needs X" is incomplete, and must be completed in the form "A needs X in order to do (or have) Y." Even what I call "intrinsic" and "original" need statements are not quite free from this inarticulate or hidden (form) condition. Somewhat, like ordinary language, needs are society-bound and also bear the stamp of time. Luxuries of the last generation are regarded as bare necessities of life by the present generation. Certain needs, such as those of food and drink, are grounded in habits and climatic conditions; certain needs are borrowed from others' (contemporary peoples) value schedules, such as those of enjoying a particular level of consumption and of performing particular types of productive work (to the exclusion of others). To make the characterization of needs more concrete, one may try to relate them to "human nature" and refer to the Marxist or the Freudian theories on the matter. But in all cases, we come across one genuine difficulty: needs are self-exceeding or dynamic both in relation to the individual and society. Even one's bodily needs, especially the quality of the goods which can satisfy those needs, undergo change and are not static. Advancement of technology and exposure to modern high-pressure publicity systems have a silent but continuing influence on our needs and values, one influencing and being influenced by the other.

All these considerations highlight the difficulties involved in the precise determination of both desert and needs, especially the latter. But this is not to deny the relevance of the concept of need to arrive at a rational criterion of justice. Relativity of needs might lead one to believe that inequality is perhaps bound to be an element of a just decision. However, that is not the case. Justice does not command that all needs of all men should be given equal weight. If some men, because of their education or imagination, can transcend the bounds of their time (history) and place (society) and present a unique value schedule, listing the goods and services that they sincerely feel should be recognized by a concerned society, I think it would not be just on the part of society to oblige them, even if it can. What justice demands is the *equality of concern:* needs of all men deserve the earnest and rational *attention* of society. When it comes to the question of *recognition* of needs, a stratified society (marked by inequality) would be well-advised to accord priority to original needs (of food and house) over imitative needs (of a colour TV set) and to original needs of the worst-off group over those of the better-off. This judgment, though not precisely formulated, seems to be quite consistent with the principle of justice. In other words, *equality of concern* and *inequality of recognition* are consistent components of a just decision. In fact, it is in-depth judgment marked by the equality of concern that enables one to realize the rationality and advisability of the inequality of recognition.

That neither the theory of social contract nor that of natural rights is out of fashion is evident from John Rawls' careful and comprehensive work, *A Theory of Justice.*[15] Instead of "the natural state of society," he uses the term "original position" to denote a hypothetical social situation in which the persons, the contracting parties, who are ordinary human beings (neither Hobbesian savages nor Rousseauist angels), are characterized by rational self-interest and concerned only with their (family's) share of the primary social goods, such as wealth, income, power, self-respect, and liberty. Moreover, they are aware of each other's capacities and limitations of the general laws of contemporary society and ethics, are "mutually disinterested," accept their position of equality, and operate behind a veil of ignorance. The original position is a *status quo* and, therefore, it is claimed, "the fundamental agreements reached in it are fair."

The concept of original position is used by Rawls in two related ways, *expository* and *justificatory.* He analyses and defines the contracting parties and conditions of the original position in such a way that he first feels justified in asserting that it is the ideal initial choice situation, and then he comes forward with the claim that the choice of the principles of justice made therein are indeed most rational and justifiable. Before we point out the weakness of this logic of choice, let us look into his formulations of the principles chosen:

> (1) Each person is to have an equal right to the most extensive total system of equal basic liberties compatible with a similar system of liberty for all, and (2) social and economic inequalities are to be arranged so that they are both: (a) to the greatest benefit of the least advantaged, consistent with the just savings principle, and (b) attached to offices and positions open to all under the conditions of fair equality of opportunity.[16]

Rawls also refers to two priority rules: Liberty can be restricted only for the sake of liberty; "the second principle of justice is lexically prior to the principle of efficiency and to that of maximizing the sum of advantages; and fair opportunity is prior to the difference principle." Lexical or serial ordering is intended to ensure that "a principle does not come into play until those previous to it are either fully met or do not apply." The object of the priority rules is to protect the interests of the least favoured, as recipients of all primary goods, liberty and opportunity, income and wealth, and the bases of self-respect. And, by implication, priority rules both pre-suppose and recognize the continued existence of hierarchical society.

Rawls' is a natural right-based theory of justice. In the original position, the basic liberties of the contracting parties are taken for granted and their rights are not subject to political bargaining or to the utilitarian calculus of social welfare. Unlike the utilitarian, he is not prepared to consider natural rights as of secondary importance, for these are unviolably based on justice and not on utility. Satisfaction of desire, as such, is of no value to the contractarian; he relates satisfaction to the question of right and also to what flows out of it. In justice as fairness, the question of good is subordinate to that of right; whatever offends or turns out to be inconsistent with the principles of justice would be disallowed. Since the contracting parties are supposed to be operating under a veil of ignorance, one does not know others' motives, propensities, and attitudes, and the consequences thereof. And that provides an additional reason for the contractarian to assert that nothing should be allowed at the initial stage that proves inconsistent with the first principles of ethics, the basis of justice as fairness. The difference between the contractarian and the utilitarian is claimed to be basic by the former; one thinks of a well-ordered society as a scheme of cooperation for mutual benefit governed by the principle of justice as fairness, the other as the efficient management of the material and human resources of society to ensure maximum aggregative satisfaction of desires as aggregated by a hypothetical *impartial spectator* on the basis of given individual systems of desires.

Rawls in not ready to compromise on the matter of principles like liberty and equality on the dubious utilitarian ground that it would result in aggregative benefit unless, of course, his priority rules are satisfied. Once the contracting parties agree to submit to the priority rules, that proves not only mutually advantageous to all concerned in their cooperative venture, but also those, who voluntarily restrict their liberties, acquire a right to a similar acquiescence on the part of those who have benefited from their submission. This is how the principles of fairness put both the benefactors and beneficiaries under reciprocal *obligations* individually. In addition, as equal moral persons, the contracting parties are required to do certain *natural duties* to support and to comply with just institutions that exist and apply to them: and since this requirement is unconditional and not concerned with voluntary acts, the related duties are called natural and thus distinguished from *obligations.*

From our limited context, Rawls' theory appears to suffer from at least one main defect. In spite of its admirable comprehensiveness, the concealed sociological assumptions and ideological leanings of Rawls' theory, when disclosed, make its application very restrictive and its claim to generality somewhat spurious.

Rawls claims that by assuming certain general desires, such as the desire for primary social goods, suitably defining the original position, and accepting the agreements arrived at that position as a basis, he has succeeded in showing how his theory of justice is independent of circumstantial relativity.[17] Again, this claim seems to me untenable.

The contractarian theory of justice is individualistic in its formulation of both the original position and the principles of justice. To minimize the effects of individualism, diverse psychological propensities, value and preference-schedules, reference has been made to an Archimedean point from which the institutional arrangement necessary for distributive justice is appraised and also to the ideal of a person whose rational plan of life and demands for goods are accepted as typical of others, and as the basis for agreements between all, leading to the principles of justice. In spite of Rawls' protestations to the contrary, it is clear that his concept of the ideal of a person is *a priori*, and his reference to Kant on this point has not proved very helpful, for Kant's concept of the rational self itself is open to the charge of a priorism. Unevenly open to the influences of social institutions, the demands for goods of different individuals are not really alike.

The ideal of a person is an oversimplifying methodological assumption. It is the concept of the good of such a person and his demands for goods that the Rawlsian theory of justice presupposes. To ensure the unanimity of choice of the individuals in the original position, Rawls substitutes for the real individual of flesh and blood an ideal, Kantian-type person. Individualism is compromised for the sake of unanimity. Then it is justified on the ground of justice, the priority of justice over efficiency and that of liberty over social and economic advantages. On scrutiny, the justification turns out to be more imaginary than real; for in the real-life situation the individuals are organized in different groups or classes and, as such, their demands for goods and plans of life prove considerably different, if not conflicting. Elements of difference and conflict are, in fact, accommodated or tolerated in the institutional arrangement and recognized from the Archimedean standpoint. That means that the scheme of cooperation envisaged in the original position seeks to legitimize social conflict and economic competition.

To meet the criticism against legitimized conflict and *competition*, the contractarian argues that they follow from the original agreements and are in accord with the principles of justice. One can be sure that individualism and unanimity do not go together. When it is claimed that they do, what is meant is either that individualism is sustained by a hidden universalism or that the unanimity principle is compromised if not spurious. Basically, this claim dilutes the value of the principles of liberty and equality.

We may note how Rawls' rider to the principle of liberty, the unequal worth of it for the people concerned, takes away the substantial value of the principle and allows its distribution as a primary good among the intended beneficiaries to be unequal. Apparently, this idea does not disturb Rawls very much, for he seems to believe that if the basic (political) liberties of the citizens (the right to vote and to be eligible for public office, freedom of speech and association, freedom of the person from arbitrary arrest, and the right to hold personal property) are established, the citizens would not be inclined to trade them off with socio-economic advantages. If the citizens were so inclined, the assumption would be that the basic rights and liberties definitive of a just society had been effectively established.

I do not question these arguments (in the form of assumptions) as such. My difficulty lies elsewhere: the effective establishment of these rights and liberties itself is contingent upon the attainment of a certain level of economic well-being. I do appreciate Rawls' giving priority to the liberty principle over the inequality principle in his *theory* of justice; but when it comes to the question of its *practice,* I do not quite see in the context of developing countries how the socio-economically worse-off people could be benefited by it in the intended manner. It seems that the developed democratic and capitalist countries of West Europe and North America, especially the United States, provide the sociological and ideological backdrop for Rawls' principles and their lexical priority. According to Rawls, the primacy of the liberty principle is non-negotiable, except for the sake of liberty itself, because "a departure from the institutions of equal liberty cannot be justified by, or compensated for, by greater social and economic advantages." My experience of the developing countries is that the very "institutions of equal liberty" are in many cases almost non-existent and the question of any departure from them does not make much sense. What is more important to note here is that the said non-existence is mainly due to the wide social and economic gap and conflict between different classes and groups. Obviously Rawls would try to meet this point of criticism by pointing out the distinction he draws between the liberty principles from the standpoint of the original position and the stage when the basic liberties have been effectively established. I note Rawls' defense of the denial of equal liberty "to raise the level of civilization so that in due course these freedoms can be enjoyed."[18]

Several questions remain to be answered. Why does Rawls propose to impose restrictions only on fundamental liberties to the exclusion of other, i.e. non-fundamental liberties, and that too the condition that the former must be legally recognized? We may note that legal recognition of what are and are not fundamental is a very shifting affair, and often the shift is brought about by and in the interests of the ruling groups. Why does Rawls propose to make unequal distribution of liberties conditional on its acceptability to those with the lesser liberty? I raise this question because the people with the lesser liberty have nothing to lose by objecting to the proposed unequal distribution of liberties unless, of course, they believe that it would further reduce their share of liberties. In the Rawlsian scheme of justice principles, this belief has hardly any place, for one must assume that the lexically prior rule, "a less extensive liberty must strengthen the total system of liberty shared by all," has already been applied. The fact that this belief has no place in the scheme does not mean that it has not been entertained. A system consisting of different and partially conflicting groups, I suspect, cannot satisfy these requirements and seems too unrealistic to be accepted by "those with the lesser liberty."

The Rawlsian scheme of social cooperation in the pursuit of justice is still unworkable in many parts of the world. This, however, is not to deny the imagination and ingenuity underlying the scheme itself. Rawls' principles of justice and the priority rules are basically *conservative* in their inspiration, designed to conserve the existing liberties of the groups and at the same time to reflect an awareness of the necessity for a just redistribution of the primary goods. The scheme of cooperation he envisages for the purpose rests on the *practical* compatibility of the first principle (of equal right of all to equal basic liberties) with the second principle (of social and economic inequalities weighted somewhat in favour of the least advantaged but qualified by the rule of unequal worth of liberties). Rawls' arguments and examples give one the distinct impression that his primary interest is to show the *theoretical* compatibility of the two principles. When facts threaten the first principle, he compromises his egalitarian position by making the worth of liberty unequal, apparently forgetting in the process the fact that inequalities in the worth of liberty are primarily due to significant second principle inequalities. No pain is spared to show that significant inequalities resulting from private ownership of the means of production are compatible with justice as fairness.

It seems to me that natural rights are not as natural as Hobbes, Locke, and Rawls try to make us believe, nor are human rights as universal as their protagonists preach, These theorists are motivated by a noble consideration—to provide a moral and an egalitarian basis for the definition and enlargement of human rights and liberties. They share an antipathy toward legal positivism, the view advocating, among other things, that the substance and scope of human rights and liberties should be determined by the laws of the land enacted by the competent authority and not left to speculative or controversial ethical principles. As we have seen, they operate at two level: first they construct their models of the state of nature or of the original position, and then derive from these the substance and scope of natural rights. Outwardly, the models appear abstract and unrelated to the socio-historical contexts in which they were conceived and constructed, but careful analysis of their elements and assumptions reveals their concealed, intended, and true character. These models are somewhat like legal fictions and serve a heuristic or an *expository* purpose. One might ask, "why they resort to model construction technique instead of description of appropriate socio-historical contexts for the purpose of derivation and definition of natural rights?" Perhaps it is because of the well-known difficulty that from the *description* of an existing situation nothing very *general,* certainly not the *ideal* principles of justice, liberty, and human rights, could be derived, and one suspects—is almost sure—that the natural right theorists had deliberately imagined the *natural* in such a way that the *ideal* could be derived from it. Whether this derivation is defensible or not is a separate question; its motivation at least is clear, and this is *justificatory.*

Justification in this type of theory is designed to work in two (often very related) ways: *direct deductive* and *inverse deductive.*

In the former case, the elements and assumptions of the model of the natural state or the original position are formulated in such a manner that the intended rights and liberties could be directly deduced from them. In the latter case, first an existing state of affairs is described, and then in order to justify it an inverse deductive reference is made to the model embodying the ideal. Any of these two methods could be used for both conservative or progressive purposes. Whereas, Hobbes uses an inverse deductive method to defend a relatively conservative theory of rights, Rawls follows a direct deductive method to defend a *relatively* egalitarian one. Neither of them seems to be interested in using a *critical deductive*[19] method, which can bring about a creative match between existing facts and ideal principles without forcing one into another and into a sort of dogmatism, or without resorting to open-ended *ad hocism*.

As we have observed, Hobbes defends only a narrow set of rights and liberties of the person, e.g. the right to live and the liberty to refuse to give self-incriminating evidence. Locke argues for a relatively broader set of rights and liberties. Mill defends equal liberties of thought, action, and expression primarily along the lines of individualism and strongly advises the state not to intervene in matters that can be managed by the citizens unless, of course, intervention becomes unavoidable in the public interest. Although these liberal thinkers—liberal in varying degrees—stand for more or less equality in the political sphere and try to justify it by contractarian or utilitarian arguments, they find nothing wrong in the socio-economic inequalities entailed by their justificatory arguments. The theoretical framework of liberalism has been designed so that political equality and socio-economic inequalities can coexist. One wonders to what extent this theoretically contemplated coexistence can be effected in practice and, assuming it can be, whether it proves peaceful. This query is very relevant, for the underlying assumption of the liberal thinkers is very questionable—that political institutions, their structure, and function can be considered in isolation from the socio-economic conditions.

Even Rawls' theory does not appear to be entirely free from this assumption. As we have observed, the special conception of justice, which, given priority of liberty over equality applies only when a society has attained a certain level of material well-being. Otherwise, in the under-developed material conditions of a society, the general conception applies and the difference (inequality) principle governs the distribution of all social goods, including liberty. It has, however, to be admitted in fairness to Rawls that in his theory at least an earnest attempt has been made to establish a weak connection between the socio-economic conditions and the realizability of equal liberty. If this attempt has not yielded a sufficiently encouraging and concrete result, it is because of the author's scheme of work itself, which has made it impossible for him to take adequate note of the diverse sociological perspectives of enactment and administration of law ensuring peace. In his understandable eagerness to construct a "moral geometry" around the concept of justice, he has resorted to an abstractionist strategy without looking into diverse, puzzling, and interesting sociological contexts of justice that perhaps might threaten his "clear and coherent" geometry. The only real society which apparently has weighed on his mind (should I say "from behind the veil of strategic ignorance"?) is understandably the American one.

I have already mentioned that to understand correctly the import of the principles of justice, and particularly those of social and economic justice, we are required to look into the sociological context in which they are discussed. The context may be considered diachronically or synchronically. To my mind, the major diachronic or historical contexts are the primitive (egalitarian) society, the feudal (hierarchical) society, the market (competitive capitalist) society, and the socialist (cooperative-conflict) society; and the major synchronic ones are the First World (one can take the United States as its paradigm), the Second World (Russia), and the Third World (India). Typological classification is bound to be approximate and therefore inexact. The resulting methodological defect can be substantially overcome by perceptive placement and measurement of the individual cases within the type. In each of the context-types there are bound to be mixed and marginal cases. For example, in the First World there are Second World and Third World characteristics: similarly in both the Second World and the Third World, one can identify the characteristics of the each. Partial heterogeneous and overlapping characteristics of the context-types suggest two apparently different, but basically consistent, phenomena: (1) principles of justice obtained in a context-type have to be lexically ordered and suitably qualified by appropriate rules so that they can prove responsive to the diverse needs of the type; and (2) every set of principles, although characteristically useful in and applicable to a particular context-type, has extra-contextual reference and significance. This is not my prescription regarding the nature and scope of the print principles of justice. This is, in fact, what we do find in the (pre-enactment) "data" of law, public opinions regarding desert, need, and contribution of the different segments of the public, moral sentiments of the people, the existing rules of social conflict and cooperation. In most cases social customs, conventions, and practices ripen into laws through legislative recognition and formulation. I hold this view because I find much substance in what sociological jurists like F.K.V. Savigny, as distinguished from positivist ones like Jeremy Bentham and John Austin, have to say on the nature of law and the principles of justice. Sociological approach need not necessarily be positive, that is, conservative and backward looking.

Admissibility of rights presupposes certain institutions and rules and also some criterion, such as desert or need, to evaluate the social goods for distribution. But the very concepts of desert, need, and social goods are differentially *used* in, say, primitive societies and market societies. In a market society a man's ownership rights to hold, enjoy, and dispose of a given piece of land, are absolute and exclusive. But in a primitive society the rights of ownership of land may belong inclusively to several persons at the same time. One may lose one's right simply because one changes one's place of residence. In a feudal society the ownership rights notionally belong to the king or, in some cases, the landlord, while the rights of possession and enjoyment actually belong to the tenant subject to his fulfilment of certain conditions. What happens in a hierarchical feudal society almost as a matter of rule is that different types of rights for different sets of people are simultaneously valid in respect to the same property. This can be found occasionally also in a primitive society. The Samoan tribal chief, for example, retains his residual right of ownership over a piece of land, which in fact is used by another person; but if the latter changes his place of residence, the right to dispose of the property reverts to the village headman.[20] This sort of hierarchical rights has outgrown both the tribal and the feudal societies and can be observed in the modern market societies as well. An owner of real estate can lease out his property to the lessee who, in turn, if not otherwise prohibited, can rent it out to the tenant; so, in relation to the same property three different rights (ownership, leasehold, and tenancy) become simultaneously valid and enjoyable. The point is what I have already said before: the analogies of a particular sort of rights found in one context-type may be discovered in other context-types, but on close examination of the contents and conditions of these rights it would be found that they are indeed *analogous* and not identical.

In primitive societies one's property rights may be over-ridden by others' individual needs or social need in general. If one is not using one's means and instruments of gathering food, others in need are free to use them as their properties. The exercise of this right is not contingent upon the owner's consent (as in the case of market societies), kindness (as in the case of feudal societies), or comradery (as in the case of socialist societies). In fact, to be helpful, kind, and even generous to the needy is a duty of a primitive man and, the former has a right to draw upon the latter's kindness or generosity. Thus in a primitive context-type it is very *natural* for the needy to tell his fellow beings, "You are bound to be kind and generous to me." But then "kind" and "generous" are being used in a different sense than ours. The "benefactor's" duties and the "beneficiary's" rights in these cases are primarily reflective of social *needs* (needs of a people living almost at a subsistence level) and not of a moral *ideal of equality* as it is ordinarily understood in a market society. Consequently, one would be advised to use such neutral expressions as the "recipient's rights" and the "giver's duties." Obviously, this game of translation of society (language) bound words and expressions which cannot be played fruitfully beyond a point.

The same point can be illustrated with reference to the issue of distribution. Primitive people follow an egalitarian practice in the distribution or divisions of the spoils of hunting and fishing expeditions. The rules governing the distribution are widely different; for example, one can deserve and get one's equal share merely by joining the expedition, by lending one's canoe or weapon for the purpose, by waiting to receive the returning expedition party on the seashore or at the outskirt of village, or simply on the ground that one needs it. This shows that the rules of distribution are not only egalitarian but also liberal. And it is very natural for tribal people haunted by constant scarcities to frame and follow the rules of desert and needs with a liberal and distinct egalitarian bias. But, I think, it would be wrong to give a utilitarian interpretation to the whole thing, for the necessary moral motivation, on which Mill would have insisted, is not there. Nor is it a matter of following the *principle* of distributive justice as fairness. From the primitive's point of view, perhaps, it is nothing more than a case of following the rule of social *practice*.

Perhaps it can be better understood in terms of face-to-face or dominantly personalized (as distinguished from dominantly institutionalized) human relationships. In this context-type familiarity, kinship, and emotional involvement mark the interpersonal relationships, whereas in the market context-type these relationships mediated by a large number of conflicting institutions tend to become more and more impersonal and anonymous. In a society that is not well institutionalized, people's rights are neither clearly defined nor easily enforceable. Endless diversity of rules and lack of definition and codification of those customary rules help the authority to settle disputes overnight, but they render the administration of justice (in the modern sense) very difficult. When a primitive man is kind or generous to a fellow man in need, it is a sort of customary or habitual response, and therefore one can characterize it as natural. When customary and habitual practices can substantially take care of disputes and conflicts, judicial authority has minimal responsibility to discharge and society is relatively homogeneous. However, this does not mean that within the primitive society one cannot draw any distinction between *moral right* and *customary right* or between *customary right* and *legal right*. Perhaps one can; but, I think, those distinctions would be quite different from our understanding of the same. In primitive society the line of distinction between practice and principles is often very thin and unclear, the former seems to be imbued with the latter. The point made earlier that natural rights, however unrealistic they might sound, have a definite *moral* undertone, lends them an air of acceptability, even if they are not immediately accepted.

Both the direct human relationships and the homogeneity that mark primitive society are largely missing in feudal society. I am talking of the Indian feudal society that existed until the eighteenth century. The main characteristic of this type of society, hierarchical order, emanated from a series of contracts, explicit or implicit. Under the king, a large number of hierarchically arranged statuses formed the pyramid of the social structure; the intermediate statuses were held by various noblemen—*subedar* (the provincial governor), *zamindar* or *bhumidar* (the big landlord), *talukdar* or *jotedar* (the small landlord); and at the bottom were the *bargadar* (the tenant), the landless peasant, and the bonded labourer. The persons of the first four categories held their statuses and enjoyed their titles as the result of explicit contract; the statuses of the persons belonging to the lowest two categories were governed by implicit contract. Each piece of land was subject to several sets of property rights and owned by the corresponding right-holders, from the tenant who actually cultivated the land, through the landlord to whom he owed service, to the landlord of that landlord, and so on. Within this general structure, the rights of a person were largely determined by custom and partly by law, and justice consisted of the protection of the rights by means of both law and custom, especially the latter. Feudal justice meant preservation of the *status quo* and therefore inequality. Attempts have been made to justify this inequality referring to the caste system and such scriptures as *Manu-Samhita* and *Dharma-sastra,* and by pointing out how the hierarchical order imparts stability to the social system, enabling it to withstand external shocks and to absorb or regulate internal convulsions. However, in spite of its inegalitarian and rigid characteristics, feudal society did allow the holders of the different statuses to establish personal relationships and to foster a sense of mutual obligations. To a very limited extent status mobility was also observed. By rendering an extraordinary service, civil or war, to the king or a superior, one could improve one's rank, add to a landed estate, or get exemption from the payment of usual revenue. But by and large, feudal society was marked by its stable hierarchy of rights and duties and highly unequal liberty.

To understand the true nature of our rights, liberties, and what is meant by justice, one has to look into the complex structure of our social system, which consists of heterogeneous subsystems (primitive, feudal, and market) and their interaction. Since we are already familiar with the main characteristics of primitive and feudal societies, let us have a quick look at the market (early capitalist) society for our limited purpose. In a real-life situation, a market society starts operating against the background of the vertically divided classes of feudal society, disturbing and dislodging the people from their fixed place in feudal hierarchy, destroying obligations of primitive kinship and traditional status. All are theoretically equal in the eye of the law, and positions are determined by contracts and exchanges, by the market forces, and not by birth or the superior's desire. Gradual breakdown of the old hierarchical structure paves the way for the rise of liberalism and individualism. Individuals are free to choose their profession, to buy and sell their produce (work and worth) in the market, to form associations and companies, to gain wealth and prestige. These market operations result in division of labour and unequal distribution of social goods, and further encourage individualism. Justification of individualism, partly a product of market society, was sought in Europe on the theory that men are born free and equal, possessing rights derived from their inherent natural capacities. Market culture and its associated ideas were brought into India and taken to the other parts of the world by the imperial powers. A new concept of justice is brought into operation by market society, and that is the requital of desert. Man makes himself, and the basis of his desert—abilities, efforts, and skills—is his own, self-formed character. Man makes himself mainly for others, and the worth of his abilities is determined by the market. The market is the ultimate desert-definer. The inequality of hierarchical society is gradually replaced by the equality of the marketplace. The process starts in the metropolitan towns, the centres of trade and industry. In the Indian context, the market society and its culture first appeared in the towns like Kolkata, Chennai and Mumbai.

While the slowly vanishing feudal features still exist in rural

India, despite all land reform acts, a new form of society—a developed capitalistic form—is steadily emerging. The workers, professionals, and administrators are becoming more and more organized through their trade unions and service associations, and have considerably improved their bargaining capacity and, in many cases, pay scales and service conditions. In contrast, the unorganized workers, especially the agricultural labourers, are very poorly paid. In their case, the minimum wage act provisions are hardly implemented; they have no leave provision; in the off-season they are literally unemployed; the talk of pensionary or old-age benefits does not make any sense to them: and, one must bear in mind, they are the single largest section of the Indian population. Of course, they *deserve* a better deal, higher wages, better service, and living conditions. Their *desert* has to be estimated in terms of the service they render to satisfy the *needs* of the society as a whole, especially those of the organized sector of the economy. When the social and economic conditions are very unequal, as they are in our society, which is in part highly organized and in part totally unorganized, equity demands that market forces should not be allowed to determine the desert of the worse-off. So in a complex society like ours, the principles of desert and need have to be suitably reconciled. But to achieve this objective the path is beset with numerous difficulties: the desert claims of the better-off are seriously objected to and at times even bitterly opposed by the worse-off, and the needs of the latter are not ordinarily appreciated or agreed to by the former. Obviously, the criteria of their judgements are not the same. Despite this, the compulsion of market society keeps them tied together by a love-and-hate relationship and engaged in a conflict-and-cooperation situation.

In the process, the market society itself undergoes change and the sub-societies tend to regroup into different and less heterogeneous classes—the working class organized under trade unions, the middle class under service associations, and the owner-entrepreneurial class under chambers of commerce. Relatively better organized and stronger groups favour the idea that the market should determine everyone's desert and that the outcome would then be fair and just, whereas relatively weaker groups want desert to be socially estimated and feel they should not be asked to sacrifice or suffer for being disadvantageously placed in a highly unequal and competitive situation.

The point that I am making diachronically, that is, by referring the questions of rights, duties, and justice to the developing and different historical types of society, can be made out synchronically as well, by referring the same questions to the different types of society of the contemporary world—developed and developing. Neither of these two types is homogeneous; for example, developed countries like the USA and the USSR, differ in many important respects; developing countries like India and China also differ on such issues as the rights and duties of the people and the desiderata of a just society.

The contemporary societies of the world have often been classified under three headings—First World (represented by the United States), Second World (represented by the USSR), and the Third World (represented by India) .[21] Making allowance for their internal unevenness, this appears to be the most satisfactory classification from our present point of view.

It is very interesting to note how human rights and justice are recognized and followed in the three different worlds and at three different levels—academic, social, and legal. It would be equally instructive to note the gap between the profession and practice of the official policies on these matters. The gap is particularly evident in what the representative states of the three worlds profess to the outside world and what they practice within their own territories. Even after the UN Declaration of Human Rights (1948), which clearly recognizes the inherent dignity, equality, and inalienable rights of all members of the human family as the foundation of freedom, justice, and peace in the world, we have witnessed, apart from the cold war between the First World and the Second World, the tragic incidents of Hungary, Czechoslovakia, and Poland, and the hot wars of Korea, Middle East, Algeria, Vietnam, Bangladesh, and Angola.

War means negation of all human rights. Even without war, human rights are often very seriously threatened. During the days of John Foster Dulles, e.g. the U.S. administration kept a suspicious eye on many leftist thinkers and activists, and some of them were harassed and persecuted. Even after the Helsinki Declaration, a number of dissident scientists and artists in the Second World have been harassed or jailed. Since the official charges against them are not open to public verification, one can hardly ascertain their substance, if any.

Although many of us are critical of the negative attitude of the rulers of the Second World toward the dissident defenders of human rights and liberties, and of the continued positive attitude of the rulers of the First World toward some dictatorial regimes, especially the military treaty allies, we cannot deny that the state of human rights in most Third World nations is indeed lamentable. In the name of national security and rapid economic development, one party rule, suppression of the opposition, preventive detention, detention without trial, and denial of justice and civil liberties have become very common.

One might say that if we accept, as I do, the contextualist theory, the above distortions of human rights and denial of justice can hardly be criticized. But this is wrong. My attempt to understand and explain rights, including human rights and justice, in terms of their sociological context should not be taken as a sort of justification. *Explanation* may be, but is not, necessarily a *justification.* When a political theorist of the First World says that traditional political and civil rights can be readily secured by legislation restraining the government's executive arm, he assumes uncritically that what is possible in a country like the United Kingdom or the United States is also possible in a country like Pakistan or the USSR. The difference in the levels of the political and civil rights actually available to the peoples of the three worlds first has to be understood and explained in terms of their different sociological contexts (including the economic, ideological, and morphological conditions). All things being equal, the countries where experiments with constitutional democratic government started in the seventeenth or eighteenth century and have struck roots are certainly placed in an advantageous position in the matter of granting civil rights to their citizens. But whether the countries of the Second World and the Third World should blindly follow the First World in this respect without assessing their own social needs and their order of priority is, of course, a separate question.

Why this question is separate becomes clear when one tries to settle the disputes between the people and the government of a developing country over such social and economic rights as "the right to work" and "the right to social security." In one sense, it seems obvious that by making suitable laws a government can secure for its citizens both civil and political rights and also social and economic rights. But before making such laws, the concerned government cannot afford to ignore practical conditions, such as social climate and executive ability, and economic resources. The government of China, for example, because of its present political system and perhaps organized public opinion, cannot allow its citizens to enjoy "the right to freedom of opinion and expression... [including] freedom to hold opinions without interference and to seek, receive and impart information and ideas through any media" (Article 19, UN Human Rights Declaration, 1948). If the government of China fails in one respect (civil and political rights) for one reason, the government of India fails in another respect (genuinely universal moral rights) for another reason. The latter, for example, simply does not have the resources that could enable it to provide each of the 700 million people of India "a standard of living adequate for the health and well being of himself and of his family, including food, clothing, housing and medical care and necessary social services, and the right to security in the event of unemployment, sickness, disability, widowhood, old age or other lack of livelihood in circumstances beyond his control" (Article 25).

In these cases, contexts are being cited as explanatory reasons for the failure of the concerned countries to conform to "a common standard of achievement for all people and all nations." The failure is not confined only to the nations of the Second World, marked by the rigours of proletarian dictatorship and one-party rule, and of the Third World, marked by poverty, low level of education and political consciousness, and different forms of authoritarianism; the failure exists also in the nations of the First World who are not only committed in principle to extend all forms of international cooperation as enjoined by several UN and UNCTAD declarations, but also are placed in the best position, technologically and financially to implement the principle. The works of economists like Raul Prebisch, Albert Hirschman,

P.K. Bardhca and Mahbab-ul Haq have shown untenability of the classical assumption that development following flow from technological innovation will *automatically* benefit industrialized and non-industrialized nations alike through the mechanism of free trade, and are also of the view that the cost factor of modernization through foreign capital might not prove increasingly differential. Under the circumstances, what is needed to establish a rational world order in which human rights can be realized and enjoyed is an economically *just form* of international cooperation; the *market form* of it will not do, for its effects are inherently differential. The existing world . "order" is patently unjust in both the Rawlsian and the Marxist senses: its inegalitarian structure, the growing gap between the poor and rich nations, is not for the ultimate benefit of the either. Growing inequalities entail even more unequal distribution of primary social goods, such as liberties and self-respect.

The structural conditions necessary for the realization of human rights are, in most nations of the world, minimal and, in some cases, almost absent. Even so we feel morally inclined to prescribe human rights for all people and all nations. Underlying this moral inclination there are two considerations.[22] First, when we make a moral judgment we make *about* something, in this case human rights, and therefore cannot logically afford to ignore the *relevant* and *genuine* (as distinguished from spurious) facts constituting the context of realizing and enjoying the concerned human rights. When we prescribe that a course of action should be followed or that some rights should be enjoyed by some persons in a particular situation, we take into account both the *abilities and desires of the concerned persons* and also the *possible states of affairs* in the situation. The relation between the two is weak but operative. The former is influenced but not determined by the latter. Factual components of a context certainly have a bearing on what (for example, a range of primary goods) the persons to a given social situation can reasonably expect and enjoy, but, I agree, there is no strong logical relation between the two. The "elastic" character of this relation becomes clear when one recalls that the factual components of a context include, among other things, common psychological and behavioural traits, customs, conventions and insitutions of certain groups of persons. Second, the elastic character of this relation between the social context and human rights of the "context-bound" persons can be used as a ground for generalizing the principles governing the realization and enjoyment of those rights. Social context performs three roles: (1) it helps us to understand how certain human rights originated in it and can reasonably be enjoyed within its boundary, (2) how further human rights, together with the existing ones, can possibly be realized and enjoyed, and (3) how human rights, realizable and enjoyable within one social context, can be gradually generalized beyond it.

To say human rights are gradually generalizable is not to say these rights are universal. A moral judgment involves a sort of evaluation of what the judgment is about: the principles used in evaluation are not the product of, but are influenced by the context in which the moral judgment is made. Undoubtedly a significant part of our dispositions, ideas, and actions have been historically influenced by our society and, at the same time, it has to be admitted to an equally significant extent that our ideas and ideals do exceed the boundary of the society we live in. It is from this perspective, one might say from the point of view of man as an ambiguous creature, that one has to understand the wide gap between the *ideals* of human rights embodied in different declarations of human rights and the *actual* state of affairs obtained in the highly uneven social contexts of the different nations of the three worlds; and one need not feel cynical about it. For our commitment to human rights is basically a moral one.[23] On the grounds that these rights remain substantially unrealized and that necessary national efforts and international cooperation are not forthcoming, we must not retreat from our commitment: we should, however, review it in the light of the disappointing but instructive experience of the last three decades.

It is evident that both the theoretical and the practical architects of human rights are very conscious of the inherent difficulties of their task. The carefully chosen language and formulations of different declarations are in effect an earnest and a continuous exercise to find "*a common standard of achievement* for all peoples and all nations to the end that every individual and every organ of society...shall *strive* by teaching and education to promote respect for these rights and freedom and by progressive measures, national and international, to secure their universal and effective recognition and observance" (UN Declaration, 1948; my emphasis). If previous strivings have not yet proved to be up to the standard and have not brought us nearer to the end, that does not mean either that the standard should be lowered or that the end should be given up. Like the natural rights defended by the classical and modern contractarians, human rights rest on some "transcendent" moral ideal that can only be partially realized in society; but, instructively enough this ideal is becoming generally recognized (at least in principle) as a basic feature of a just world order. Gradually, this recognition of the ideal as a principle is likely to percolate down to the level of practice, lead toward positive legislation and executive implementation. But, then, by that time the ideal of human rights itself will be further broadened, deepened, and redefined. In the pursuit of human rights and justice, perhaps we are well advised to avoid the extremes of positivism and romanticism or utopianism.

REFERENCES

1. Thomas Hobbes, *Leviathan,* introduction by A.D. Lindsay, E.P. Dutton, New York, 1953, ch. 13, p. 66.
2. *Ibid.*, ch. 14, p. 66.
3. *Ibid.*, pp. 66-67.
4. John Locke, *Two Treatises on Government,* ed. Peter Laslett, Cambridge University Press, Cambridge, 1960, p. 75.
5. *Ibid.*, pp. 322-23.
6. *Ibid.*, pp. 58-59 (author's italics).
7. *Ibid.*, pp. 325-26.
8. *Ibid.*, pp. 303-08.
9. Brian Barry, *Political Arguments,* Humanities Press, London, 1965. ch. 3.
10. David Miller, *Social Justice,* Clarendon Press, London, 1976 pp., 22-23
11. John Stuart Mill, *Utilitarianism: On Liberty; Representative Government,* ed A.D. Lindsay, London, 1964, p. 46.
12. Henry Sidgwick. *History of Ethics,* Macmillan, London, 1967, p. 284.
13. John Rawls, *A Theory of Justice,* Harvard University Press, Cambridge, Mass. 1973. sect. 48.
14. Barry, *Political Arguments,* ch. 5.
15. Rawls, *A Theory of Justice,* pp. 11, 28, 32.
16. *Ibid.*, p. 302.
17. *Ibid.*, p. 263.
18. *Ibid.*, pp. 61ff, 54243.

19. D.P. Chattopadhyaya, *Individuals and Societies: A Methodological Inquiry,* 2d ed., Scientific Book Agency, Calcutta, 1975, ch. 5.
20. Margaret Mead, *Cooperation and Competition among Primitive Peoples,* Beacon Press, Boston, 1961, p. 292. See also A.S. Diamond, *Primitive Law: Past and Present,* Methuen, London 1971; and M. Gluckman, *Politics, Law, and Ritual in Tribal Society*, Blackwell, Oxford, 1965.
21. Irving Louis Horowitz, *Three Worlds of Development,* Oxford University Press, New York, 1972.
22. R.M. Hare, *Freedom and Reason,* Oxford University Press, New York, 1972, see also C. I. Lewis, *The Analysis of Knowledge and Valuation,* La Sall, Open Court Publishing Co., 1946
23. D.P. Chattopadhyaya, "Freedom and Human Unity: Ideal and Actual," in *History, Society and Polity,* Macmillan Co., Delhi 1976, pp. 234-61.

14

Law, Morality and Polity

I

The relation between law and morality is controversial and is being debated over the centuries right from the time of Kautilya and Plato, if not earlier. The question of relation between law and morality, on the one hand, and polity, on the other, has remained open to debate for an almost equal length of time. One of the questions that has to be settled is whether law is independent of morality. The question may be put in another form: Is legal responsibility altogether different from moral responsibility? Sometimes, this question has been formulated in a more specific form: Is there any cut-off line in between individual responsibility, which *is essentially moral,* and collective responsibility, which is said to be *essentially legal?*

If law could be rightly deemed as secular, i.e. independent of morality, then one could perhaps plausibly argue that polity has nothing essential to do with morality. It is often assumed that the main defining characteristic of law is its *enforcibility* and not (what is considered) its *moral sanction.* Admittedly, it makes hardly any sense to speak of self-enforcing morality, unless, of course, self is taken as a metaphysical norm and/or endowed with a native capacity to transform its will (which is required to be absolutely rational, i.e. non-hedonic and non-somatic into a law binding on itself and on all other rational selves). If law without moral sanction has to be accepted as binding on all rational selves, then excessive demands are being implicitly made on the human rationality, or on the authority of the polity, or both.

Problems tend to be acute when it is agrued that human beings in general are not evenly rational : that most of us at times behave in a manner which cannot be regarded as rational or even reasonable; and that only few of us are steadily rational in all our behaviours. To save a rationalist theory of morality one, of course, can dismiss these matters as merely factual as if these matters were not of much consequence in settling the concerned moral questions. But, then, it may be pointed out, this sharp fact/value dichotomy thesis has its own difficulties, especially in the realm of law: for the enforcibility of law in a particular (factual) situation is a constituent character of law itself (which is supposedly value laden).

If the moral sanction behind law is denied or withdrawn, then the latter's enforcibility tends to be an exclusive jurisdiction of the polity or the civil society or the State authority. In that case either polity itself can be deemed to be based on a moral basis, or, denying it a moral basis, sufficient authority may be given to it on the grounds of *praxis* and *needs.* The former alternative was accepted by the followers of Hegel, and the latter one by those of Marx.

To a man of common sense, law is a *necessity* for maintaining an order (or a hierarchy of orders) in the relations obtaining between the human beings of a particular society and also for governing the transactions which take place between them and their associations. To a man of common-sense it is not patent that this necessity of order is an expression of an underlying freedom. The inter-human relations and transactions constitute the basic structure of polity. To make these structures possible and durable, it seems to me, both *necessity* and *freedom* have to be there. On the basis of pure needs we could not possibly form a polity. For its survival is contingent upon the willing consent of its constituents to some regulations mutually agreed upon or externally suggested or imposed. The human beings have some *moral sentiments* to come and stay together under an authority of their own make, which, once made, is allowed to conduct its affairs with relative autonomy. This enables authority to govern the inter-human relations and transactions normally without hindrance ordinarily posed by the momentary impulses, moods and capricious actions of the concerned humans. This, however, does not mean that the authority of the polity is insulated from questions and corrections grounded in human reason, needs and freedom.

II

Justice is the key concept which brings out clearly the intimate relation between the concept of law and that of morality in the context of polity. To ensure that inter-human relations and transactions become and remain just, what the concerned beings and their associations are called upon are : (a) to obey the rule of law, and (b) to submit to the legitimate coercion resorted to by the lawenforcing authority. This entails a double responsibility on the part of every citizen: (i) the responsibility of upholding those of, the said relations and transactions which are just, and (ii) the responsibility of forbearing so that the same responsibility may be discharged by others. Justice is the need of the free people to preserve their freedom for fulfilling their needs, natural and creative. Main among the natural needs are the biological ones like freedom from hunger, satisfaction of sexual urge and protection against weather hazards. Prominent among the creative needs one may count development of science, art, and technology and all that which enables human beings to increase their understanding of and control over the forces of nature and the area of their mutual cooperation, minimizing the basic causes of conflict and gradually removing the non-basic ones.

The theory of justice that I propose to defend is ability-based and need-oriented. Here "and" has to be taken seriously: for it is indicative of the close relation between a historical concept of freedom and the moral desirability of its need-oriented application. The proposed theory of justice is inconsistent with what I call the structural theory of justice propounded, among others, by Plato and also with the quasi-structural one defended by thinkers like Kant, Rawls and Strawson. I am not quite sure in which of these two camps Kant should be placed. For some of his basic ideas have been ingeniously used by the quasi-structuralists like Rawls and Strawson who, it seems to me, are not as pronouncedly anti-historical as Plato and Kant. But, again, one may point out that the Plato of the *Republic,* often cited as a proponent of political authoritarianism,[1] is quite different from the Plato of the *Laws,* a proclaimed defender of the rule of law.[2] Mine is not primarily an exegetical study but a conceptual one.

Plato is credited with the authorship of two different but isomorphic concepts of justice—psychological and social. The psychological concept of justice states that an individual is just if each of the three parts of his soul (the reasoning, spirited and appetitive) functions, optimally, resulting in that state of inner peace, amity, and concord which may be called psychic harmony. The social concept of justice states that a polity is just if and only if the concerned men employ their individual abilities more to the needs of others and less to their own. Both in the soul and polity the harmony is hierarchic. In the polity the hierachy consists of the guardians, the citizens and the slaves, in that order. The division of labour and production for the market are said to be the generative principle of the polity.[3]

Every man is called upon to follow that line of *social conduct* which accords best with his natural endowments and acquired skills, for that alone, Plato thinks, contributes maximally to the happiness and welfare of the polity. The emphasis is on "social conduct". The psychic harmony of the individual by itself is not enough.[4] It must be evident in a stable and consistent manner in his actions towards and transactions with others, which could be accepted as an indirect proof of his dispositional orderliness and predictability. Plato's insistence on the desirability of establishing and maintaining structural unity between the hierarchic harmony in the individual's soul and that in the polity should not detract one's attention from the fact that he wants to see that a polity is ordered, reflecting the orderliness of the soul. The harmony in social conduct is contingent upon the psychic harmony. The privileged status of the latter is not sought to be compromised by Plato when he speaks of the desirability of harmony or orderliness in the socio-political, i. e. external, sphere.[5] On the contrary, it seems that his intention is to try to shape a polity as a prototype of the harmonious soul which itself is said to be energized by the contemplation and the *vision* of the hierarchical world of Ideas. Social justice is, then, a manifestation of the just state of man's soul which, in turn, mirrors in its pure state what "really is", the eternal world of Ideas. The pure soul of the philosopher can "see" with its "eye" justice itself (an Idea) and be just, and also knows best what a just polity would be like.[6] What a man does, how he conducts his affairs with others, is expressive of his soul. Justice is good in and of itself (for the just man) because it makes the just man happy. The just man is not covetous and does not seek for self-advantages. To be just, a man must follow his own nature, abilities and inclinations.

This is, in brief, a structural theory of justice; "structural" as distinguished from "historical". The world of Ideas, the vision of which energizes the human souls in different ways is changeless. The changeless world of Ideas (of Justice itself, Beauty, etc.) is viewed by the philosophers, the ordinary citizens, and the slaves with different degrees of distinctness and clarity, depending on the purity (or lack thereof) of their souls. Those who have a clear vision of the world of Ideas are internally better organized, i.e. their traits and dispositions more stably or durably harmonized.

Structurally, a comparable theory of justice, which is ontologically less committed and therefore formally more neat, is found in the writings of Kant. It is on the united will of the people that he thinks that all rights and justice rest. And this unity is due to some *a priori* laws of reason[7] to which all conduct of rational human beings is subject. A proposition about rights (or justice) with respect to an *actual* object is analytic: for other's interference with it adversely affects my freedom, unless, of course, it is presupposed that the interferer has a legal right (a right superior to actual possessory right) to what I actually possess (infringing upon his freedom). To possible objects, i.e. those which are not actually possessed by this or that rational human being at some particular time and place (or other), rights extend by virtue of some *a priori* principles of the form: "it is a duty of justice to act toward others so that external objects (usable objects) can also become someone's (property)."

Legal possession is conceived by Kant as independent of actual possession(s) and yet making the same possible (i.e. intelligible) legally. Just ownership or right is a *de jure* union of the will of a person with an object, not necessarily requiring actual/physical possession. This union is possible (i.e. intelligible) only in a juridical condition of society, under a public-legislative authority, i.e. a civil society. In a natural state of society, possessory rights (or justice) is *provisional de jure* which, in civil society, matures into a *peremptory* character. It is the union of will of all expressed in the form of law that puts all subjects under the obligation to honour the possessor, rights of each.

Kant draws no sharp line of division between *natural justice* and *legal justice*, the former is regarded as the inarticulate form of the latter, and what unites the two is presupposed freedom. Consequently, in the concept of justice, *obligation* is more important than *enforcibility.* But both are required to ensure the possibility, actuality and necessity of the possession of objects in a civil society. Protective justice makes it *possible* for man to have meaningful right to an object. *Actual enjoyment* of that right is taken care of by reciprocally acquisitive justice. Distributive legal justice provides the conditions necessary for the possession of objects in accordance with laws.

Kant's polity, like Plato's, is an Idea (not to be taken in the ordinary English sense of "idea"), an archetype, *a priori* and changeless. This is supposed to provide "internal guide and norm" for every possible union of men in the civil state. The legislative authority, based on the united will of the people, is the fountain of all that is right and just, and, therefore, "can do absolutely no injustice to anybody."[8] Proceeds from this "holy and inviolable" authority is so binding on all that, Kant says, "it is a crime even to doubt it or to suspend it for an instant". Like Plato, he also denies the very possibility of a right of sedition, "much less a right of revolution." Going further ahead, Kant asserts in a stern *a priori* voice: "....it is the people's duty to endure even the most intolerable abuse of supreme authority." His "rational" interpretation that "all ruling power comes from God" (i e. "some highest perfect legislator") is, contrary to Popper's claim,[9] hardly liberal. He is totally against questioning the supreme authority of the law-giver, and the only reforms he stands for are those which the latter himself brings out voluntarily. Kant's argument, though does not sound convincing at all to me, is very clear.

> The Idea of a political constitution in general is holy and irresistible, [for] it is an Idea that is an absolute command of practical reason judging in accordance with concepts of justice—a command binding on every people. Even if the organization of the state is defective by itself, still no subordinate authority in that state can bring any active resistance against the legislative chief of that state. Rather, the deficiencies that are attributed to him must be gradually removed by reforms, which he carries out by himself. Otherwise, if a subject were to adopt a conflicting maxim

> (or proceed in accordance with his arbitrary will), a good constitution would come into being only as a result of blind chance.[10]

Kant's formulation of "the Idea of a political constitution" and its interpretation as "an absolute command of practical reason" heavily draw upon Plato's theory of Ideas[11] and tend to rule out the very possibility of genuine battle of ideas and ideals, without which a civil society is sure to lose what may be called its moral and democratic content. For any political experiment or programme or whatever is *a priori* morally impermissible under the Idea of a political constitution, cannot be lawfully undertaken by anyone without exposing oneself to Kant's stricture against "the vulgar appeal to so-called adverse experience." If the whole of mankind is to operate for all time to come and under all circumstances with a unique archetype of polity, then any attempt to establish any other type of polity is bound to be frowned upon and thought to have been founded on a *conflicting* maxim. And such an attempt, however ideologically motivated it might be, is declared irrational and illegitimate. Once an abstract and seemingly eternal idea of polity is allowed to hold its sway over the human thought and action, then *genuinely* new political attempts, possibility of creative response to historical challenges of poverty, inequality and war, are *a priori* ruled out or, even if grudgingly tolerated for the time being, unequivocally condemned. The structural theory of justice and its variations are basically anti-historical, anti-creative and conservative. Even Kant's is no exception.

What follows from the structuralist theory of polity and justice is most likely to offend one's historical sense and common sense. How, for example, could one be reasonably expected to accept with equanimity a lawyer-king or a military king or for that matter, even a philosopher-king who turns out to be a tyrant or an oppressor? How, in the face of repeated acts of tyranny of the ruler, one could be reasonably expected to believe that he is the fountain of justice and, therefore, (by definition) can do no wrong? Have we not seen the rise of many tyrannical regimes under the guise of constitutional government in the different parts of the world, developed and developing, in the last sixty years or so? Have we not seen so many cases of visible "constitutional" emasculation of constitution with or without the backing of "convenient" judicial interpretation of law and also the cases of silent killing of laws, especially the "inconvenient" ones, without administering them properly? The poor who needs and often also deserves most the benefit of law has to remain content only with its penal discipline and restriction, while its fruits are available only to those who can pay for the high price of justice. Yet the structuralist's advice to the poor and the rich is identical (obviously ignoring the difference in their position in life): "follow law and your own nature, and do not try to reform it, for that is inconsistent with the basic character of justice and would harm you."

III

The counterpart of the structuralist in practice is, the strict constructionist, who, in the words of Lord Denning, "still holds, [the] fortress," insists on following the precedents, and impedes the growth of justice through decided case-laws. Lord Denning is a liberal, not revolutionary, certainly not against the doctrine of precedent.

> ...not...that I am against the doctrine of precedent... All that I am against is its too rigid application—a rigidity which insists that a bad precedent must necessarily be followed. I would treat it as you would a path through the woods. You must follow it certainly so as to reach your end. But you must not let the path become too overgrown. You must cut out the dead wood and trim off the side branches, else you will find yourself lost in thickets and brambles. My plea is to keep the path of justice clear of obstructions which would impede it.[12]

I share Lord Denning's concern for keeping the path of justice clear of the obstructions created by the law-court followers of the structural theorists of justice. I am against the latter and the strict Constructionist, for both of them impede the growth of justice, think that their voices are the "voices of infallibility," and, thus, encourage a sort of moral laziness. I endorse the liberal constructionist's enterprise "to cut out the dead wood and trim off the side branches." But one is advised to remember that in the structuralist wood of justice, nothing dies, because nothing lives, and that there are no side branches, because growth is unknown in that eternal garden of Eden. Whatever is traditional can survive the onslaught of time and the (*sham*) test of (*shadowy*) historical experiences. And we are told, whatever is traditional is rational and just, and must not be tinkered or tampered with by the fallible creatures.

I am aware that I may be accused of lacking in objectivity in my understanding of the structuralist theory of justice. I have already tried to argue, though not elaborately, why have I come to this view. However, I am aware that there is restatement of the structuralist position, which, in the absence of a better expression, I have called the quasi-structuralist theory of justice. And the latter, it seems to me, is different from the former only peripherally and not in substance. The quasi-structuralist tries to save the structuralist position against the charge of anti-historism or growthlessness by trying to bring it closer to the concepts of moral sense, moral sentiment or conscience of the common man of all times and places. "Times and places" enter into this theory as "shadowy things," and, therefore addition of "all" does not add to the strength of it. Moreover, the "common man" of "all times and places" is a mere abstraction somewhat like Plato's and Kant's "essentially rational man". I say "somewhat" because the quasi-structuralist's essentialism is attenuated to lend a historical touch to it. While the structuralist lays emphasis on the *eternity* of values, the quasi-structuralist on the *stability* of the same. "Stable" values, it appears, do not owe their origin to history, but survive through it. Among the main ingredients of this theory one might count, (a) stability of values, (b) their apprehension by moral sense or moral sentiment, (c) their relevance to, and yet substantial independence from, all times and places, and human beings, and (d) the inherence of values in human actions. The said ingredients could be found in the writings of such diverse thinkers as Cudworth, Shaftesbury, Hutcheson, Whitehead and Strawson.[13]

Cudworth, often regarded as the most systematic metaphysician of the Cambridge Platonist school, thinks that all that is *active* is spiritual and none that is *passive.* Only when mind by its active nature has itself as its object, knowledge is possible, knowledge of eternal truths. Neither a human sovereign nor even God can make an act good merely by willing it: either it is eternally true or eternally false, that acts of certain kind are good. Only when we act out of love, not out of sense of duty, we rise above egoism, become free, disinterested and objective, and enter into harmonious and concordant relation with other human beings. Shaftesbury is for unfettered freedom. The only cure of wrong reasoning is more reasoning, free reasoning. All understanding proceeds from *self* understanding. Hobbesian atomism and egoism are inconsistent with the human nature and the benevolent nature of God who orders the world. Men are disposed to act virtuously because of their inherent moral sense which discloses their unity and common interest in the public good. Moral sense is said to be like a sense of harmony in music and a sense of proportion in art and architecture. Hutcheson develops Shaftesbury's theory of moral sense along the lines of Locke's psychology. Reflection on moral sense makes us aware of the original nature of human actions and of their two related characteristics, virtue and benevolence. Moral knowledge is not directly due to God; but indirectly due to God that all men have this discerning faculty in them. Like sight, the moral sense is universal in mankind, enabling it to apprehend immediately the light (or virtue) of benevolence and the darkness (or wrongness) of violence (against other persons). But some men have defective sight or are born blind; and these men tend to do wrong things and indulge in acts of violence. The moral sense is endowed not only with the *simple* ideas of what is right and what is wrong but also those of approbation or condemnation of the same.

What makes me think that Whitehead is very close to the moral sense school, apart from their common Platonic affiliation and accent on the concepts of good and activity, is his simultaneous recognition of *process* and *pattern* or structure in reality. Unlike many other quasi-structuralists, his familiarity with history is very scholarly and profound, and had elicited reverential reference from his friend and student, Russell, himself a prohistorist. By "value" Whitehead means "intrinsic reality of an event". Every event functions under some pattern; but pattern itself does not ensure the realization of an event, i.e. the attainment of value. The most general ontological condition necessary (but not sufficient) for the realization of value is pattern and activity.

> The notion of the importance of pattern is as old as civilization. Every art is founded on the study of pattern. Also the *cohesion* of social systems depends on the *maintenance* of patterns of behaviour; and *advances* in civilization depend on the fortunate *modification* of such behaviour patterns. Thus the *infusion* of pattern into natural occurrences, and the *stability* of such patterns, and the modification of such patterns, is the necessary condition for the *realization* of the Good (emphasis mine).

Both ontology and sociology of change have been ostensibly recognized by Whitehead. To reconcile "stability" and "modification", he speaks of "adjustment" and "infusion", and, it is interesting to note, in the definitions of these key concepts, psychology is scrupulously avoided. Even in the definition of *beauty* as "the perfection of harmony", and in that of *harmony* as "the perfection of Subjective Form", one finds no psychological term. Like "adjustment", another teleological concept significantly used to reconcile "modification" and "pattern" is "adaptation". Like most other quasi-structuralists, Whitehead hates "discord" because of its alleged "destructive and evil" character, welcomes "harmony" and orderly social change. His accent on "stability" is also indicative of the author's otherwise known conservative cast of mind.

Strawson's formulation of the defence of the quasi-structuralist position, in spite of its careful, cautious and perceptive character, hardly conceals its basic orientation.[14] He says, "it is a pity that the talk of the moral sentiments has fallen out of favour." By "moral sentiments", he means, "the general structure or web of human attitudes," which is said to have in it "endless room for modification, redirection, criticism, and justification." But "the general framework" is autonomous and which "we are given with the fact of human society." Strawson seems to reject both : (a) the "one-eyed utilitarian's" optimism and determinism based on calculability of the consequences of human actions, and (b) the libertarian's pessimism based on the *prima facie* realization of the gap between the description of a situation and attributability of moral responsibility to an agent (working in it) for his attitudes and actions and, then, his plan "to plug the gap with an intuition of fittingness." Both (i) the incalculability of all consequences of what we do, and (ii) the gap between the description of a situation and what actions are morally warranted in it, i.e. the problem of fittingness, it is claimed, can be taken care of in terms of Strawson's theory of *internally* improvable moral sentiments. These sentiments and attitudes are said to be impervious to external, i.e. socio-historical influences. On the contrary, the latter are "manifestations" of the former. "Our practices do not merely exploit our nature, they express them." By providing enough room for improvement, refinement and modification of human nature from within, Strawson seems to think that he has succeeded simultaneously in sobering down the utilitarian optimism, by bringing it closer to what happens in history, i.e. we abandon those of our practices which prove to be wrong or unworkable, and also in warding off "the obscure and panicky metaphysics of libertarianism" which relies so heavily on "intuition of fittingness" and facile "repeated verification".

Scrutiny of his writings reveals that Strawson is against historical and anthropological approaches to understanding the principles of morality and law. Rather, consistently with his thesis of the moral sentiments and their autonomy, he prefers a psychological approach, which backed by a regressive-analytic method, enables him to *discover* that "general structure of human attitudes and feelings", which underlie permanently below the diverse cultural settings often studied by the historian and the anthropologist. This is further evident from his strong disapproval of Collingwood's attempt to reconstruct historically Kant's *a priori* categorical structure of understanding and also of Körner's welcome idea to replace Kant's doctrine of space and time as *a priori* particulars by relatively dynamic notion of space-time as suggested by Whitehead. Search for relatively stable framework of thought and action tends to make the quasi-structuralist partly blind to the history of the framework itself. For he thinks that history owes its own intelligibility to and within a given framework. Haunted by the pseudo-problems of unity and continuity of human thought and practices, one gets the impression, Strawson wants to arrest the course of history. Lest it strikes against the framework from without and breaks it up, he wants to see history at rest (Plato's divine) or at least relatively immobilized *within* the framework.

While criticizing Levi-Strauss's structuralism, Sartre traces its origin to what he calls "tyranny of analytic reason" and shows how dialectical reason and history can well take care of the unity and continuity of human thought and practices.[15] In his eagerness to preserve the "unity" of the structure, the structuralist denies the reality of time and change and the quasi-structuralist, displaying more perception of realism, tries to bring and contain "continuity" within it. I have argued the point at length elsewhere.[16] One historical unit may be dialectically de-united and then re-united, preserving some of its elements and eliminating others, and, thus, accounting for both continuity and unity of our thought and practices.

1V

It seems to me that there is no unique just polity which can perfectly or even reasonably meet the needs of all peoples living heterogeneously in separate groups and classes in different societies. There is no universal form of human justice. And this becomes increasingly clear when we move from the sphere of municipal law to that of international law, when we discuss the state of human rights in the different countries of the world and even after so many declarations of the international community,[17] how wars have been conducted and prisoners of war treated in the last seventy years or so, and what steps are being taken to establish a just and equitable economic world order. I do not want to say we should not try to evolve common norm of actions and homogeneous living. Certainly, all of us should. But this is a moral ideal and not a description of what most of us are doing. When an ideal is prescribed for all, knowing fully well it is unacceptable, for some compelling or not so compelling reasons, to most of us, then it is really meant for few, and, therefore, it is hardly anything more than an appeal and lacks in legal force. Neither law nor morality is autonomous and has any social force-free internal structure of its own: both dialectically interact with the less articulate and more earth-bound structures of biological and social needs. When some laws, though based on otherwise laudable principles of morality, are of no relevance to the needs of the poor and the weak in a particular society, lie dead in the book of statutes, for these do not command common respect and are not followed. True, law cannot respect everybody's whim, for in that case, it cannot rule; and that is not my point. But if law disregards interests and needs of a sizeable section of the people, it loses its moral worth and becomes ordinarily unenforceable; and even then if it is enforced, it amounts to illegitimate coercion or violence. In exceptional cases the rich may be victim of violence; ordinarily, almost constitutionally, it is the poor. The poor needs the help of law most, and often deserves it also, but rarely gets it; for the price of administered justice proves prohibitive for him, particularly in the context of a developing country like ours.

An unfortunate upshot of the structuralist theory is denial of the variety of what is Good, and of moral and political development. For, Goodness, according to this view, is *essentially* the same everywhere (a unique, simple and indefinable quality), and its so-called variety *accidental* and, rightly understood, lacks in moral import. What this line of argument entails is: morality and also polity, marked by variety or development has to be construed as something a-moral. Certainly, it is not the intended import. A good government, in that case, means *an indefinable quality of excellence* in the acts of governance which need not (or, perhaps, should not) aim at political or economic growth, removal of exploitative conflicts, and just integration of conflicting groups. Even Rawls' theory of justice, otherwise a serious enterprise carried out under the mixed influence of Kant and Sidgwick, suffers from the defects of structural rigidity and stratification. His *special* conception of justice as fairness, which accords priority to equality over liberty, becomes operative only when a certain (indefinite) level of material welfare has been achieved in a society. Otherwise, the *general* conception operates and the difference principle is allowed to govern the distribution of all social goods, including liberty. It has been rightly pointed out that while Rawls seeks to ensure equal liberty to all, his effort seems to suffer from the questionable assumption that the worth of liberty is not same to different classes in society.[18]

The structuralist theory of law, morality and polity seems to be founded upon an "indefinable" and, therefore, indefinite notion of freedom. Freedom is said to be equal liberties to all. But, on own admission of the structuralist and the quasi-structuralist, this is a presupposed ideal and not a practically available situation. Different theorists of the said persuasions are inclined to accept the pre-supposed (pre-analytic or intuited) freedom also as the main political and (post-analytic and to be achieved), disregarding, interestingly enough, all existing and glaring social differences. The presentation of "the soul-polity isomorphism" thesis, preserving the division both in the soul and also in the polity intact, provides an interesting reading. For while it pleads for "harmony" in the soul, it is not clear how the division in the soul could be consistent with the "aimed at" harmony. The same question is pertinent in the sphere of polity as well : how a class-ridden society could be made "stable"—stability, a desideratum of good society, and also geared upto do its best to realize "the common end", assuming there is one?

A satisfactory answer to this sort of questions, may, perhaps, be plausibly constructed, provided we are prepared to give up the *a priori* notion of freedom, i.e. an ability innately given, in favour of a notion which defines freedom partly as an *ability* obtained in the normal body-mind complex and partly as a consciously entertained need-fulfilling, and constraints-removing *objective.* By implication, this notion of freedom is historical and contextual, but not autonomous.

Freedom is not a matter of natural or spiritual inheritance; it is to be achieved working hard on what is below, around, within and beyond us. The theory of "internal" improvability and refinement of our dispositions and attitudes ignores the dialectical interactions which unavoidably, often educatively, take place between "the external" and "the internal" human factors.

Once we recognize this important practico-dialectical aspect of freedom, I think, we would be in a position to identify our partly peculiar and partly general *needs*—moral, legal and political, and to work, as far as possible together, for the removal of those needs.

Neither abilities nor needs, unless very abstractly defined, are universal or a-historical. The moral and political needs of a developed and a developing polity, for example, are not the same, neither priority-wise nor, at times, even item-wise. What a good polity in India would be like? It is sort of question which cannot be meaningfully discussed, still less decided, unless, given the heterogeneous groups and conflicting interests at work, we succeed in identifying our needs and abilities. And that brings us in touch with the down-to-earth reality, viz. causes and cures of poverty, problems facing judiciary at different levels, multiple and conflicting professional morality, and the willing or the unwilling default of the intellectual,[19] and so on. At this stage, philosophy, if it were to deliver any good, should transform itself into a praxis.[20]

REFERENCES

1. Karl Popper, *Open Society and Its Enemies* Vol. 1, RKP, London, 1974.
2. Glenn R. Morrow, "Plato and the Rule of Law", in Gregory Vlastos (ed.), *Plato 11, 1971.*
3. *Republic,* (369Aff).
4. *Republic, Bk IV* (444 *C-E).*
5. *Republic, (433* A-B).
6. J. Walter Jones, *The Law and the Legal Theory of the Greeks,* Oxford 1956 p. 116ff. See also *Republic (5001).*
7. Kant, *The Metaphysical Elements of Justice,* tr. J. Ladd, Bobbs-Merrill, New York, *p. 57.*
g. *Ibid.*, pp. 77-78; see also, pp. 84-86.
9. Karl Popper, *op. cit.*, p. 247.
10. Kant, *op. cit.*, p. 140.
11. Kant, *Critique of Pure Reason, A* 315-320, B372-77.
12. Lord Denning, *The Discipline of Law,* Butterwortbs, London, 1979, p. 314.
13. J.A. Passmore, *Ralph Cudworth,* Cambridge, 1951; R.L. Brett, *The Third Earl of Shaftesbury,* Hutchison London, 1951 ; D. D. Raphael, *The Moral Sense,* Oxford, 1947; La A.N. Whitehead, 'Mathematics and the Good" in P.A. Schilpp (ed.) *The Philosophy of Alfred North Whitehead,* Open Court, Salle. 1971 ; *Science and the Modern World,* New York, 1925; *Process and Reality,* New York, 1929.
14. F.A. Strawson, "Freedom and resentment" and "Social morality and individual ideal," in *Freedom and Resentment, and Other Essays,* Methuen, London, 1974, *The Bounds of Sense,* Methuen, London 1966. p. 44. pp. 118-21; see also, Stuart Hampshore (ed.), *Public and Private Morality,* Cambridge Univ. Press 1979.
15, Jean-Paul Sartre, *Critique of Dialectical Reason,* tr. Alan Sheridan Smith, New Left Books, London. 1976, pp 800-05, 817-18, 823-24; Claude Levi-

Strauss, *Structural Anthropology* I, tr. C. Jacobson and B G. Schoef, Allen Lane (Penguin), London, 1969, ch. 1; *Cultural Anthropology* 2, tr. M. Layton, Penguin, 1973. ch. XVIII.

16. D.P. Chattopadhyaya, *History, Society and Polity*, Macmillan, New Delhi. 1976, pp. 197-219; see also, my *Individuals and Worlds: Essays in Anthropological Rationalism*, Oxford University Press, New Delhi, 1976, pp. 170-208.
17. D.P. Chattopadhyaya, "Human rights, justice and social context," in *The Philosophy of Human Rights,* (ed.) Alan S. Rosenbaum, Greenwood Press, Westport. Connecticut, 1980; (Ch. 14 of this book).
18. John Rawls, *A Theory of Justice,* Oxford, 1973, Sect. 11, 26, 82; see also, Norman Daniels (ed.) *Reading Rawls: Critical Studies of A Theory of Justice,* Blackwell, London, 1975, pp. 253-81; Ronald Dworkin, *Taking Rights Seriously*, Harvard Univ. Press., Camb., 1978, ch. 6.
19. D.P. Chattopadhyaya, "The philosopher and the intellectual default," in *Dialectics and Humanism,* No. 3-4, 1976.
20. Jurgen Habermas, *Theory and Practice,* Heinemann, London,1974, ch.7.

15

EPILOGUE

I
VALUE AND NATURE

Value-seeking is a part of human nature. That man is said to be a uniquely *self-exceeding* creature is mainly because of this value-orientation. His biological past and sociological environment, though necessary, are not sufficient to explain his *drive* towards and *need* for new values. Being conscious, at least partly, of the operation of the laws of nature and culture, man is in a position to *assimilate* the given and *anticipate* what is not yet given. His power of anticipation works in respect of both facts and values. And this power itself, as I have said before, is, to a great extent, native to his nature—a legacy of evolution and environment. Strictly speaking, it is only as an integral part of a given segment of nature and a particular society that man understands himself, intends and acts to fulfil certain needs. In other words, man does not understand, intend or act as an abstract or unrelated self. In order to vindicate the immutable character of values some thinkers, taking their cues of thought from Plato and Kant, have conceived the *solitude* of man as the paradigm of his freedom. The *unrelated* man is, in a sense, free, but that freedom does not enable him to tell what is good and right for other persons and places. Unless the "free" man is found to be free and cognitively competent enough to speak unerringly of others' values, to legislate for and on behalf of others, his freedom, by definition, turns out to be sham. The fear of nature either because of its "tooth-and claw" (Neitzsche)[1] or because of its "causally determined" character (Kant)[2] is responsible for the definitions of freedom in terms of unrelatedness or solitude, withdrawal from nature and culture. Interestingly enough the defenders of these notions of freedom ascribe to the withdrawn, i.e. free man, the competence to tell us which values are universal and necessary, i.e. not related to and conditioned by the considerations of particular persons and places. The said fear of nature rests on a serious misunderstanding of the relation between man and nature, an untenable dualism. The same misunderstanding and the resulting fear form the basis of such questions as "how can values be there in a world of facts?", and dualistic or proto-dualistic answers to such questions, saving the "universal and immutable" characters of values from causal encroachment.

II
MEANINGS AND DEFINITIONS OF VALUE

Before we indicate how values are shaped by environment and evolution, I think, it would be useful if we say few words in order to clarify what is meant by "value". Plato,[3] Moore,[4] and Ross, among others, have convinced us of the difficulties involved in trying to define such ethical notions as "justice", "right" and "good". Kant argued why nothing else than "good will" should be regarded as "unconditionally good". With them the basic ethical notions are *intuitive,* immediate, and cannot be obtained in and through any thing else. So they find it possible to delink values from facts, refuse to trace the influence of the latter on the former, and assert that the concepts of environment and evolution have nothing to do with values. According to them, values almost *unilaterally* apply to or are expressed in, but not influenced by, facts. But, faced with such commonsense questions as "Don't we have *historical* accounts of moral values in different cultures?", the defenders of permanent values feel uneasy, but dismiss the same by saying "you are confusing the questions of principles or definitions with those of their cases of application or satisfaction" or some such thing. Even those who recognize, as we do, the distinction between *principle* and *application* feel that the relation between the two is not one-sided and that the importance of *application* must not be berated. For, in that case, one fears, values tend to become matter of theoretical legislation or definition on the authority of unquestionable intuition or conscience.

Practical inadequacy or uselessness of the legislative or

definitional approaches to the problems of values have, at times, inspired another opposite, i.e. empirical survey, approaches. It has often been seen that when we, placed in a conflict situation, consult some specialist known to have studied the related sort of situations carefully, and ask for their suggestion on possible ways out, they calmly tell us, "that's not our task; we are not practitioners in the sense lawyers, physicians and psychiatrists are." This is perhaps partly correct as a matter of tradition. And many thinkers are of the view that this tradition, though majestic in its uncompromising autonomy, should be abandoned and gradually replaced by values as practised by different groups of human beings, including the groups of the defenders of autonomous and permanent values. The dichotomy between theory and practice, definition and application/illustration, has unnecessarily made some otherwise important theoretical discourses, especially the moral ones, irrelevant to the various pressing issues of life, thought and action. And hence, primarily as a reaction to this *practically indifferent theories of values,* some philosophers, working in close collaboration with experimental psychologists, biologists, and sociologists, have tried to study values from empirical scientific points of view. In this connection, the names which readily come to my mind are those of R. B. Perry,[5] S. C. Pepper,[6] W. Kohler,[7] E. C. Tolman,[8] G. Allport,[9] C. I. Lewis[10] and Charles Morris.[11] Some thinkers, influenced by psychoanalysis, existentialism and Marxism, have also made significant contribution to the field of practical dynamics of values.

Perry defines value *descriptively* in terms of *interest* and in such a manner that the traditional objection against the so-called dualism between affection and conation cannot be raised against it. Anything has value or is valuable if it is the object of interest. And our interest is a train of events determined by expectation of its outcome. It is to be noted here that the "train of events" has psychological, physiological and physical components in it. The expected object of interest induces actions for realizing or preventing it. Both Tolman and Perry offer a purposive theory of value; while the former's main interest is behaviouristic animal psychology, the latter's interest in psychology is neither exclusively behaviouristic nor primarily concerned with animal "appetites" and "avoidance". Yet their main views are mutually complementary.

We humans, like animals, are "drawn towards" the objects which can satisfy our "appetites" of thirst, hunger, sex, and play, and want to "avoid" cold, heat, and obstruction. The question arises whether humans are *driven by* their appetites or feel *drawn towards* the objects which are *known* to be capable of satisfying the same. Primary drives of animals and humans are not identical in extension or intension. The gregariousness of animals, for example, is not what we understand by our social life. In the matters of exchange, interaction and organization, it seems, human organism is relatively better equipped. The fact that we do not feel causally *pushed* even by our primary drives or *irresistably* pulled by the objects *known* to be capable of satisfying our appetites shows one thing very clearly and, that is, we can, at least partly or temporarily, arrest or hold back our drives, allowing our knowledge or cognitive memory to play a role in weighing/evaluating the possible means leading to the goal and the outcome of achieving the goal itself. Induced by appetites or needs when we are aware of our goals and feel like getting to the same, we can afford to wait and ponder, may be with a quiescent tension within. Once this intermediate stage between the primary drives and the actions for realizing the goal is recognized, I do not see how we can deny that our active consciousness or cognition has a *say* in shaping the scope and strength of the primary drives themselves. Given that *say*, it appears that the primary drive of all adult humans or even of children are not identical.

The point I am trying to make is that between appetition or drive, on the one hand, and conation, activities directed to the realization or the avoidance of the goal, on the other, there is an intermediate biocognitive stage. It keeps the concerned organism informed of the functions necessary to get to the goal. It is this power of anticipation which makes it possible for the organism to recognize the means and adapt itself to the same, facilitating the realization of its goal. What is true of the human organism is also true, to a lesser extent, of the animal organism. For man can better classify, relate, retain and, on demand, make use of the results of the previous drives, anticipations and efforts. *Relative* inability of the animal organism to invent and use signs necessary for the purposes of *regulation* of drives, anticipation of and *adaptation* with the means (related to its goal) and the resulting handicap have been partially compensated by nature endowing it with the competence to be in *immediate* touch with the encoded *signals* of the environment and the biological advantages to act accordingly in the pursuit of goals (avoiding dangers). Man's *abstractive* capacities have made him a citizen of a big civilized world but, at the same time, alienated him, to a great extent, from the closeness, richness and freshness of nature, making him more vulnerable to the new and hitherto unknown forces of nature. Aware of this limitation of his endowments, man has been making increased use of such mechanical aids as microscopes and telescopes, to extend his sensory discriminations. To extend his motor coordinations he is using telephones, radios, televisions, etc. By clever use of language mathematics, logic and computers he extends his relational discriminations. But the question remains: are these devices and their ramifications being used to get man rid of his biological legacy, or to lift him above the forces of nature? One suspects that a large section of the representatives of what is known as scientific culture is *objectively,* i.e. may not deliberately, promoting a worldview in which control and successful use of nature are being systematically highlighted as the necessary conditions for the progress of the human race. One wonders whether by forgetting or insufficiently recognizing the biological foundation of cognition and the natural basis of culture, we can better define our relationship with our environment or, to put it in another way, carve out a better position for ourselves and our descendents on this planet.

Whether we follow or repudiate Mill's *please-pain* principle or Pepper's *achievement-aversion* principle, it is difficult to deny man's *bipolar* relation to and position in nature. Whether we accept or reject Perry's *concept of interest* of Kohler's *concept of requiredness,* one thing seems obvious: the concept of value cannot be adequately clarified and brought closer to our persistent moral experience without challenging its *autonomy* claims. The static-structuralist's concept of autonomous and timeless values rests on a certain distrust towards the laws and the evolutionary processes of nature. It is also partly responsible for his *formalistic* method of discussing the moral issues, somewhat in disregard to the moral experiences persistently reported by different peoples. His preoccupation with *normative ethics* and *forms* of moral judgments rather than with the practised mores or ethos and the stuff of moral experience is also partly explainable in terms of his ahistorical attitude. The sharp distinction drawn between *normative, ethics* and *practical ethics* is not only uncalled for but also untenable. It makes checking of norms impossible. It is a fallacious corollary of the stuff-structure dualism defended, among others, by Plato and Kant. In contrast I refer specifically to the names of Mill, Perry, Pepper, Lewis and Kohler because it seems to me that whatever difference might be there between themselves, they have perceptively tried to close up the supposed gap between nature and culture, between "is" and "ought", taking the former as their basic point and tracing its empirically ascertainable relation with the latter. Many of us forget that Mill, often accused—perhaps wrongly—of committing naturalistic fallacy, was clearly aware of the distinction between value-in-use and value-in-exchange. And if he insisted on drawing the desirable, value, from the desired. psychological fact, it is primarily because of his empirical monism belief in the underlying unity of the realms of values and facts. Perry acknowledges his debt to Mill and defends the latter's theory of "right" as conduciveness to moral good and "wrong" as conduciveness to moral evil[12] Pepper's appreciation of Perry's works are unmistakable,[13] Kohler's commitment to gestaltism and phenomenology did not prevent him to commend the worth of Perry's work and his main work on ethics has been dedicated to Perry. Lewis like Perry, Pepper and Kohler, is clearly opposed to value/fact dualism and also thought/action dualism. Knowledge involves valuation and valuation needs knowledge. Thought itself is a form of action and thoughtless action is either simple reaction or, at least, non-human. It is unfortunate that "thought" and "knowledge" have been defined by some thinkers as timeless, impersonal entities primarily to ensure their immutability and objectivity.[14] Without correct evolutionary perspective, the world of ontology becomes unnecessarily overpopulated, and frozen and proves to be at variance with our experience.

III
AGAINST VALUE-FACT DUALISM

The most promising aspect of the studies in values by Perry, Pepper,

Kohler and Lewis is their common acceptance of the broad perspective of evolution and sociology. Each of them, except Kohler, was deeply influenced by Dewey. The same may be said of Charles Morris's studies. Instead of confining their attention to the narrow base of language analysis or theory of knowledge, they have taken the broad perspective of evolution and within which both language and knowledge figure with added significance. For example, the vexatious problem of the *definition* of values has been avoided by Morris by adopting a more promising line of inquiry, that is, by letting the peoples of different countries to define and rate their own values. To identify the *determinants* of values and, if possible, quantify, their weightage in relation to individuals and groups is perhaps more important task than to try to get hold of the elusive *defining* properties of values. It is true that before one accepts this as task one has to give up the indefensible idea of autonomous value. Only this minimum point has to be recognized because human organism "becomes a self through participation in social interaction, the self obviously is influenced both by its organic substructure and by the society in which it develops."[15] It is no wonder that the defender of autonomous values in order to discount or minimize the biological and sociological influences on man's moral life is often obliged to fall back upon the idea of the immortality of soul. The *eternal* values theoretically need *immortal souls* to sustain them.

I for one do recognize, though uncomfortably, the *theoretical* considerations leading to the *postulation* of eternal values and immortal souls. I am more uncomfortable with those who claim that eternal values and immortal souls are objects of *cognition* and not postulation. But here I do not propose to enter into those larger, though, very related, issues. Once it is shown that values are not subject to the process of time, the findings of biology, sociology and other empirical disciplines can be ruled out from the so-called autonomous domain and declared irrelevant to the definition or determination of values. Because of its changing characters, if the individual human life or even the social life cannot be regarded as the *locus* of value, then it has to be ascribed to the eternal soul as its native quality/competence. It is recognized that souls are embodied, at least sometimes, that bodies act or are moved by the unmoved soul(s), and that *somehow* at the level of action, bodily passions and social influences make their presence felt. It is in this context one can clearly understand why the classical rationalists and their modern followers systematically highlight the advisability of following the light of reason or will only to the extent it is rational, i.e. free from the influence of impulse, passion and desire. Unable to recognize actions in their passionless or disembodied purity, the rationalist concentrates his attention on moral standards, ideals, or norms with reference to which the moral worth of actions is to be judged. And thus arises the uncalled for distinction between normative ethics and practical ethics. I am not against drawing this line of distinction for purely analytic and clarificatory purposes. But the theoretical motive underlying the drawing of this line of distinction becomes clear when we are told that these norms are permanent, necessary for estimating the moral worth of our actions, and it makes no sense to speak of their changeability. Yes, for the purpose of estimating and rating the moral worth or appropriateness of actions in a given context, we do need and make use of some "moral" criterion or standard, but that need not be and in fact is not permanent. The choice of criterion itself, consistency or perfection, e.g , presupposes some sort of familiarity with and valuation of the objects or facts—actions, attitudes, policies, etc. to which it is intended to be used. Borderline cases or odd instances are always there necessitating the review, if not redefinition, of the criterion itself. It is somewhat like the problems of taxonomy in zoology and botany. Evidently, these issues are somehow rooted in stuff/ structure or matter/form dualism.

IV
FOR VALUE-FACT ISOMORPHISM

With reference to values, Kohler seems to have made a very significant study to do away with the stated forms of dualism and paving the way for a sort of phenomenological gradualism. Kohler's study is all the more important because he is both a structuralist and an evolutionist. Structuralism, broadly speaking, may be of two types, Platonic and Kantian, objective and subjective. Neither the Platonist nor the Kantian would be so uncritical as to deny that the world of experienceable facts and that of values are related. The question is: "how?" Plato's answer is that our moral values and whatever has moral value are affiliated to and reflection of *absolute* values in another world. That world itself knows no change and therefore can explain the unity and continuity of the changing experienceable facts. Values, to be recognized as such, are *required* to be reflective or imitative of the absolute values of *that* world. Kant also speaks of an absolute value, duty, but discovers its source in the absolutely rational will, or unconditionally good will, of the human soul. The right thing for us to do is what is our duty and the formal structure of that duty, correctly understood, makes the modes of discharging duties of all rational beings *consistent* or harmonious. Our awareness of what is our duty could be called correct or authentic only when it is in harmony with the awareness of duty of all other men of goodwill. In brief, the awareness of duty to be recognized as truly moral is *required* to follow from the universal formal structure of the soul and thus to be in harmony with the awareness of duty of all men of rational will or good will.

Kohler complains that it is not at all clear what Plato and Kant mean when they insist that our moral experiences, experiences of facts recognized as valued by us, are *required* to be reflective of *eternal* values of a *transcendental* world or to follow from a universal formal structure of the soul. Their eagerness to prove eternity, universality, objectivity and harmony of values is understandable, i.e. anti-sceptical and anti-relativistic but, Kohler feels, their arguments do violence to the nature of "perishable facts" and laws of science, and are unconvincing. In order to find a satisfactory relation between the perishable facts and their changing laws, on the one hand, and psychosomatic moral experiences, on the other, he turns to the phenemenological method of Husserl and makes extensive and intensive use of the concept of *requiredness.* The concept of requiredness is bound to remind one of Husserl's concept of *intentionality* and Perry's *interest. Interest* sounds subjective to Kohler's ear and to bring it closer to the *field* of facts and make it compatible with the same, he introduces the concepts of *directedness* and *vector.* In fairness to Perry it has to be admitted that he is never tired of emphasizing the role of *object* of interest in his analysis of "interest". The concepts of directedness and vector highlight two aspects of a moral situation which, it is alleged, have not been adequately recognized by some pragmatists: (a) subjective phenomena like interest, drive, desire, etc. are neither discrete nor random in the field of their mutual relation, and (b) they exhibit some properties in their behaviour indicating their requiredness even in relation to the facts of the objective world (physical and biological). Occasionally when Kohler speaks of "incompleteness", it reminds one of the property of objectwardness or out-thereness of the intentionality of the mental phenomena. But requiredness, as it has been observed, is not peculiar to the mental or psychical phenomena but exhibited also by the facts studied by different physical and life sciences. In the absence of this property, one can easily imagine, facts can never coexist with and be satisfactorily responsive to the value-oriented phenomena of the psychical field. Self is also recognized as a source of vector. But it would be wrong to think that the directedness of the psychical facts is entirely due to the self: for the latter itself is a phenomenon among phenomena, of course, with a distinction of its own and as a relatively enduring point of reference. By his phenomenological analysis Kohler wants to dispense with the notions both of "psychical causation" and "physical causation." And he brings to our attention three main points : (i) psychical facts and forces form a structure, (ii) physical facts and forces form a structure, and (iii) these two structures, taken *macroscopically*, are similar, i.e. *isomorphic*. Requiredness is expressive of the psychophysical *isomorphism* and confirmed by the experiences of satisfaction, fulfilment, and fittingness, for example. *Isomorphism* of the neurobiological space and physical space have to be understood *functionally* and *dynamically.*

It is interesting to note that to show how values and facts can live together in the same world, Kohler first points out that they as *phenomena* are *isomorphic* and mutually *required* for each other, and, then, adds that this relation is *dynamic* and evident at, the *macroscopic* level. Among the views he repudiates, are, (a) the *isomorphism* between a *microscopic* fact and "its" psychical or linguistic counterpart, (b) the *static* character, (c) the *ineffability* of the isomorphism, and (d) the "logically" unbridgable gulf between facts and values. Interestingly enough, all the four repudiated views, (a) to (d), are intimately associated with the name of Wittgenstein and on whom Kant's dualistic influence, as pointed out by Stenius, is considerable. Kohler does not criticize Wittgenstein referring him by name. During 1937-38, when he was writing his book, *The Place of Value in a World of Facts,* Wittgenstein's view on the *awkward* position of values in the world of "facts" became quite influential through his followers' writings. Carnap, Ayer and Stevenson were arguing then that values factually are rootless and express only our attitudes, likings, etc.[16] It is true that, of late, emotivism makes indirect allowance for cognitivism, is tempered by naturalism, and the perceptive ones among the ethical thinkers have been trying to trace and clarify the complex connections between our value statements and facts.[17] But elements of dualism or tension remain.

V
COMMON PARENTAGE OF FACTS AND VALUES

Value-fact dualism does not pose a unique problem. This dualism is contingent upon how one defines these two concepts. If the concept of force is adequately recognized and allowed to have its say in the formation and transformation of both facts and values, the theoretically raised barrier between them breaks down. "Force", "power", "energy" and similar other terms, though used interchangeably in ordinary language, have their definite meanings in physics, biology and other sciences. *Generally* speaking, we are obliged to accept and make use of a concept which can take care of "things" describable by such terms as "moving", "changing", "passing", "in process", etc. Often the term *motion* is called upon to serve the myriad purposes. In the absence of such a concept, not only dualistic but also monistic systems, spiritualist and materialist ones alike, are clearly in difficulty. Without admitting the existence of the *power* of Becoming in Being spiritual monists, the Parmenidean and the Vedantin, for example, find it difficult to explain the world of multiplicity and value. Somewhat similarly, from the other end, materialist monists, Democritus and Marx, for example, can hardly account for the motion, combination, evolutionary change, and qualitative enrichment of the ultimate material particles without postulating the existence of motion-*in*-matter. It is an insistent verdict of human experience that both facts and values are in the world and that they are dynamically related. If the dualist's ontology does not permit him to be fair to the insistent verdict of his experience, he has to modify it by adding auxiliary ontological entities or epistemological devices or both. Similar strategy is resorted to by the monist as well. In the process, such concepts as "entelechy", "emergence", "life-force", "nodal point", "organismic". "intentionality", "requiredness", "dialectics", "prototeleology", etc.' gain *theoretically necessary* currency.

If Aristotle is liked by many biologists and evolutionists, it is not merely because of his experiments in marine biology, which are undoubtedly important, but perhaps primarily because of his evolutionary monistic ontology which was formulated with an expressed axiological orientation in view, identifying God with Good, and thus tracing the parentage both of facts and values to the highest Reality, the "Unmoved Mover". The adjective "Unmoved" is said to be a Platonic hang-over on Aristotle's mind. That which makes movement possible but does not share *in any way* the motion of the moved or is not affected by it *in the least* is bound to be considered ontologically quite independent of the latter; and, in that case, the problems of dualism reappear. Using this argument from down below it has been reasoned that if the physical facts and forces had nothing in common with the facts and forces of *life,* the evolution or emergence of *life on the earth* would not have been possible. It is difficult to deny that there are striking similarities between the behaviours of the inorganic matters and those of the organic ones. At times this has induced some molecular biologists to proclaim that the only worthwhile way to study life is to be carried on in terms of its chemical and physical components and processes. One can trace the motive-force behind this reductionist approach to Descartes who, deeply impressed by the mechanist hypothesis of his time and of which he himself was a main propounder, thought that the laws which govern the motion of a watch or other automation and those which can explain the motion of an animal body are identical at bottom. Darwin disputed the claim that all biological explanations and laws can be deduced from the basic physico-chemical laws and, simultaneously, pointed out their intimate connection, continuity and interconnection. That laws of biological evolution constitute a science on their own right was a revolutionary claim clearly made and convincingly established by Darwin. The leading biologists of the modern time do endorse this basic point of Darwin, may be in different ways. And, interestingly enough, the gain of biology, i.e. the vindication of its claim as an autonomous discipline, proved to be loss for ethics, at least for those who had been claiming that values are autonomous and have nothing to learn from or do with biology.

VI
EVOLUTION AND VALUE

All of us, including the non-human creatures, are *objectively* concerned with values. The needs of life and living oblige us to seek certain values and avoid or confront certain disvalues. It is indeed very gratifying to note that a large number of life scientists have been taking increasingly informed interest in the problems of value posed by evolution and its various ramifications. Perhaps their intimate understanding of the complex aspects of the process of living and the causes of life and death, of the origin and extinction of species, has something to do with it. The critical conditions which now environ our life, both individual and collective, on this planet have drawn their concerned attention. When I am saying this I have in mind particularly such biologists of distinction and with larger sympathies as Huxley.[18] Haldane,[19] Dobzhansky,[20] Dubos,[21] Simpson[22] and Bronowski.[23]

To be aware of what is value is a part of our nature. Genetic cravings, biological competence and environmental informarion are at the roots of our value-seeking thoughts and actions. The ideas which are being sharply focussed and experimentally shown by the working biologists were not entirely unknown in the past (Dubos: 21, 1975). Living in the lap of nature and, simultaneously sustained and threatened by its forces, man just could not afford to be indifferent to the ways of his interaction with nature. Endowed with the capacities to learn from experience, retain the outcome of learning in the categorized manner, and transmit it down the descendants, the species-man acquired, through the process of evolution, a clearer understanding of his relation with his own past, with the changing environment around him, and of the workings of nature within himself. In the ancient and "sacred" books of the east and the west on cosmology and medicine, e.g., the discerning reader can easily find out the perceptive understanding of the intimate and complex relation between man and nature. Many of the ancient legal, moral, medical and even administrative systems speak of "the wisdom of body", "the common root of orderliness of all things terrestrial and celestial", "the accumulated wisdom of generations", "the advisability of following the laws of nature" and of kindred concepts. Both Hume and Adam Smith held that moral sense arose in *natural* manner and their repeated reference to the roles of bodily organs and stimuli provided by environment is also interesting. But, needless to add, in the absence of the evolutionary context of biology, they could not tell us definitely of the operational mechanism used by nature to *direct* the course of evolution and of the varied responses of man. Now we are authoritatively told:

> A man consists of some seven octillion (7×10^{27}) atoms, grouped in about ten trillion (10^{13}) cells. This agglomeration of cells and atoms has some astounding properties; it is alive, feels joy and suffering, discriminates between beauty and ugliness, and distinguishes good from evil. There are many other living agglomerations of atoms, belonging to at least two million, possibly twice that many, biological species. What is most remarkable is that the individuals of everyone of these species are so designed that they are able to live and reproduce in some existing environments.[24]

This wonderful state of living affairs has been shaped during billions of years of evolution. And the principle which has made it possible is natural selection. It is the differential reproduction of the carriers of different hereditary endowments. Heredity is a conservative force and confers stability upon biological systems. Evolution is its antithesis. If heredity can work perfectly, evolution is not possible. Genetic variation brings about diversity between individuals or groups within a species, or the diversity between species. As a process it means that the development of individuals is modifiable by environmental forces, and that the hereditary endowment is changeable by gene recombination or mutation. It is because of the accumulation and ordering of mutational changes that all organic diversity comes out of the monophyletic origin, the unity of the primordial life. In brief, evolution, marked by genetic variation and open to the forces of a given environment, works as the principle of individuation in the life-world entailing the preservation of an increasing number of species and the extinction of some species. Though heredity is said to be a stability principle, it is dynamic stability. Even genotypic, not to speak of phenotypic, stability is analogical and relative, and not exact. The environment in relation to which heredity works consists of many changing systems of organic and inorganic matters and is itself dynamic in character.

That man has been able to assume the best and the highest position among all other living creatures is undisputed. He could adapt himself best to the dynamic conditions of life. Making ingenious use of his hereditary endowments and the environmental resources he alone, of all mammals, has been able to develop culture, which, unlike the biological properties of life, can be lifted above the relatively fixed hereditary lines and transmitted and expanded freely. By culture, man has increased his control over other living species and nature. Culture is nature transformed by man using his nature-given competence. He has reason to be proud of his achievement and to feel dismayed at some philospher's observation that the finest flowers of his garden named values have a dangerous pest in them called "naturalistic fallacy". We are told by Simpson[25] that we need not bother about the fallacy of "naturalistic" ethics or that of supernaturally sanctioned "counter-naturalistic" ethics because in either case the autonomy and authority of man is being compromised. In the positive part of his theory he consoles us by reminding that normal humans do have a moral sense, can decide between right and wrong, better and worse, and act appropriately. "It is thus plausible and indeed, I think, practically certain that ethicizing the capacity and the necessity for some system of ethics, arose of human evolution in a completely natural way." Man's ethical consciousness, one may say, is the combined outcome of (a) genetic capacity to learn and transmit, and (b) environmental support frequently offered in the form of challenge demanding of man to exercise his capacity in new and diverse ways. With Simpson and Huxley, Dobzhansky[26] shares the view that ethics is evolutionary and in a way peculiarly human; but, they add, evolution does not give us an ethics or a set of values which can be readily used in different situations. It is for us humans, to develop and change our values. Evolution does not have a purpose of its own. Having denied God's purpose in evolution and ethics, they do not find any good reason to credit the process of evolution with any purpose and which could be regarded as foundation of human ethics. It is very true that evolution provides man the *background, competence* and *environment* which are all necessary for man to know nature, develop culture, and moral consciousness. But, it is to be noted, man does not derive his morality either from God or from evolution. He is a moral agent without any Principal to control him and to whom he can refer back and turn round in need. And that makes his moral consciousness all the more self-accountable, deep and serious. He can look back to nature for minimal support and to culture for some guidance but finally has to fall back upon his own human resources (not directly and causally sustained by either of the parent sources). Man's moral responsibility is therefore immense. And in a different but serious sense, it is antonomous.

Strictly speaking, the expression "evolutionary ethics", without clarification, is open to misunderstanding. Evolution of nature by itself does not tell us how we humans should think, formulate our norms of action, and act. Even if it is assumed that through its course of evolution nature has progressed and developed, we do not decide or know the goal or goals towards which it is progressing. May be we can anticipate and influence it partially. Even if we could know it more adequately, it is not at all certain, whether we would have accepted it as our goal. If under nature's compulsion we would have accepted, then that "acceptance" would have been without much moral significance. To speak of "acceptance" of nature's unknown goal as our own goal makes little moral sense.

But this is not to deny the more important point already made out earlier, viz., the course of evolution, particularly its upper biological stage, left certain cues which have proved of immense importance for anthropoids to lift themselves gradually above the causal "compulsion" of nature, develop culture, and broaden his area of influence through successful use of tools, signs and symbols. Waddington[27] going a step further claims to have found strong evidence to believe that even sub-human species of animals do have their moral consciousness and act accordingly. He thinks that the existence of ethical beliefs is a necessary part of the evolutionary system and that once we recognize this fact, the so-called circularity of evolutionary ethics allegedly evident in Julian Huxley's theory, viz., naturalistic ethics can neither be *independent* of nor *derived* from evolution, does disappear. On the basis of well established facts and one's own observation of animal behaviours, one feels strongly inclined to endorse the view that ethical impulses and judgments are there among the animals. The problem of circularity seems to me somewhat misconceived and therefore the question of solving it does not arise. Rightly understood, "naturalistic ethics" is neither *absolutely* independent of nor strictly derived from evolution. Causal laws are influential but not compulsive.

As I have said earlier, "naturalistic ethics" (or, should we say, human ethics) is *relatively* independent of and weakly derivable from certain characteristics of evolution—adaptation, fittingness, organization, integration and differentiation, etc. I wonder if the sufficient condition requirement of defining "absolute independence" and "strict derivation" can at all be satisfied in this biological evolutionary context. It seems to me that organisms, encounter with their environment and generate in them certain properties which gradually enable them to encounter the same environment more to their advantage and change it, and that, in the process, certain other properties *gradually* emerge in them increasing and refining their capacities to define more successfully their relation between themselves and their changing environment and that this process, marked by variation, regress, and progress, continues. But the continuance of the process is not a matter of course or law: it is contingent upon so many conditions, internal and external; and that accounts for the extinction of many species. The point I am making is this: once we accept the view that sense of value *gradually emerges* in us, evolves through stages from the animal levels to proto-anthropological and anthropological levels, the question of circularity does not arise. Gradually emerged out of nature, and educated by evolution both genetically and environmentally, we have been able to form culture, and values symbolize its forwardly engaging network of relations and thrust.

If Waddington has taken us to the animal world to show the convincing ethical sense of the otherwise feared, avoided and disliked, animals, Julian Huxley[28] takes us to a newvalue-based religion, evolutionary humanism. His concept of religion is naturalistic, based on the discoveries of physiology, biology, and psychology and which not only make it possible but "necessitate" it. Man needs religious beliefs to deal with his destiny, to adapt himself to the new situation, containing both hopes and dangers, a mixed bag of the intended and unintended results of his knowledge, technology and action. Dangers are high energy weaponry systems, thoughtless urbanization, mindless exploitation of natural resources, senseless use of drugs and surgical instruments, and of inhuman pollution of the environment. Hopes are man's love for truth and knowledge, desire to develop himself, and concern for others. Like Huxley, many of us think that these hopes are not operative at their optimal level. Man can "reconcile his self-fulfilment with others" need for the same. The ideal of "comprehensive wholeness" is realizable. "Collective awareness is... the distinctive and most important organ of the human species. It can be improved both quantitatively, by adding to knowledge and extending the range of experience, and qualitatively, by improving its organization." Julian Huxley's idea and language betray his welcome optimism. He thinks that evolutionary humanism "both necessitates and makes possible the reconciliation of extreme positions and the adjustment of conflicting interests". Possibilities of growth are open. It is for man to make judicious use of them. Huxley refers to Charles Morris's concept of "the open self". His sympathy for the view of Pierre Teilhard de Chardin is well known. The catholicity of his mind is remarkable. One only feels that he has not sufficiently highlighted the negative aspects of the evolved man of today and tomorrow.

Besides Huxley and Waddington, J.B.S. Haldane has also

closely studied the issues related with man's evolutionary past and future. But left to himself, he is not very interested in passing judgment of value on evolution, for he thinks that we do not know enough about it. Nor his language, in spite of its wit and elegance, is very reverential while referring to "progress", "purpose". and "the future man". Without ceasing to be scientifically serious Haldane can be devastating in his expression.

> I have been using such words as "progress", "advance", and "degeneration"... but I am well aware that such terminology represents rather a tendency of man to pat himself on the back than any clear scientific thinking. The change from monkey to man might well seem a change for the worse to a monkey. But it might also seem so to an angel. The monkey is quite a satisfactory animal. Man of today is probably an extremely primitive and imperfect type of rational being. He is a worse animal than monkey. His erect posture leads to all sorts of mechanical troubles, such as hernia and a narrowing of the pelvis which makes childbirth painful and dangerous. The last stage in man's evolution certainly has its dark side.[29]

Mind, according to Haldane, has played very little part in evolution. Study of cerebral biology suggests that it is possible to interpret certain systems of material objects as *mindlike* and also to interpret the conscious behaviour of man *mechanically.* And he sees no reason why this should horrify Bernard Shaw or thinkers of similar persuasions. It is difficult to imagine why Shaw's ancients in *Back to Methuselah* and H. G. Wells's *Men like Gods* like humans of today, would not be subject to certain chemical and biological laws, maybe of broader scope making wider variation possible.

> If human evolution continued in the same direction as in the immediate past, the superman of the future would develop more slowly than we, and be teachable for longer.... He would probably be more intelligent than we, but distinctly less staid and solemn.

The most important *value* that Haldane can derive from his study of evolution is *beauty* presented by the laws of the universe. The scientist's concern should be to unveil "the austerely beautiful outline of reality", which is often veiled by colourful imagination of man, the "theorizing animal". Like Spinoza and Einstein, Haldane commends "the truly scientific attitude. . .a passionate attachment to reality as such, whether it is bright or dark, mysterious or intelligible." Having commended this well-known scientific, i.e. value-neutral attitude, he himself admits that "no scientist can constantly preserve it." But why? Serious attempts to answer this question start unveiling the irremoveable value dimension of man like ethical animal.

VII
ENVIRONMENT AND VALUE

Maybe it is true, and perhaps it is, that till now the present mind has not shaped the course of evolution significantly. But from the records of the last few thousand years, not a very long period in the time-scale of biological evolution, it appears that mind has been increasingly asserting its role in its relation to its earthly environment, both natural and cultural. It has been possible mainly because of man's deeper understanding of the laws which govern the concerned facts and forces. "Understanding", "knowledge" and similar other terms, as we have noted before, have an evaluative element in their defined meanings. Whether man's increasing control over his environment is an unmixed good is now an open question. We need not enter here into such questions as whether, in fact, "man's control over the environment is real or imaginary" or "whether the increasing control as such is good or bad." For, while I admit that the said control is neither decisive nor univocal, it is diffcult to deny that in the last hundred thousand years or so man's ability to manage, utilize, and contain his environment has increased. Besides, the talk of "control as such" is somewhat pointless because what is meant by "control", or for that matter "knowledge", never works as such, but only in conjunction with a cluster of related concepts, some of which are emotive and evaluative.

By his very nature man seems to be not only "ethicizing" and "theorizing" but also ,"sign-using" and "symbol-using". Foot-firm in nature, these capacities of man have now made him a citizen of three worlds—the natural, the mental, and the relatively transcendental. Those who reject oceanic monism, pluralism, and sharp dualism have been speaking along this line for a long time. Evolutionism has provided scientifically plausible links connecting these worlds. In the recent years, Popper and Eccles have presented a very well argued theory of three worlds.[30] The world-1 contains, among other things, (i) matter and energy of cosmos, (ii) structures and functions of living beings, and human brains, and (iii) such artefacts as machines, tools, books, works of art, and music. The world-2 consists of subjective consciousness, perception, imagination, emotions, memories, dispositions, etc.; and the world-3 contains the records of intellectual works, theories, arguments, problems, etc.[31] It is to be noted that unlike Plato, Popper speaks of *interaction* both between world-l and world-2 and between world-2 and world-3, and, therefore, is obliged to believe in a sort of evolutionary continuity between them. I take *continuity* as a weaker form of unity. The strong claim of unity, methodological or ontological, is an upshot of one or other type of *reductive monism*—spiritualist, materialist, or vitalist, and seems to be at variance with the qualitatively widely different facts and forces discovered by scientists. These worlds are *gradually* related and their relation is due to the dialectics of evolution. Man is perhaps the middle term of this evolutionary process, as Sri Aurobindo puts it. The end of this evolution, if any, is not known.

What is being deeply felt is that the problems or issues raised hitherto evolved three worlds, have put mankind under very complex, enormous, and at times almost unbearable, strains and conflict. As a race, man has not been able to manage well the forces he, aided by nature and culture, has unleashed. His knowledge of the laws governing the facts and forces of three worlds is not proving adequate for him to define a satisfactory relation between himself and the environment. Gradually he is coming to know of the sources of enormous powers, energy and competence within and around him, but, curiously enough, the systems of society and values built by him and his fellow beings over the years are being found to be at conflict with the end of realizing those powers. His hard-won freedom of knowledge brings him into painful touch with the fetters of action.

Genetically, man remains relatively stable. Genotypic form of reactions of man to his environment remains more or less steady over generations in spite of various genetic mutations brought about by biological, chemical, and environmental factors. Our cultural environment and, in many cases, physical environments, as we know, are fast changing because of various factors—industrialization, urbanization, mobility, deforestation, high energy experiments, extensive use of synthetic products, medicines, etc. Germplasms are not easily and directly modifiable by the environmental influences. Under the circumstances, the problem man is increasingly facing is how to adapt himself to the fast *changing* environment with a *stable* or *conservatic* genetic apparatus. True, adapting he is. But the biological and the psychological costs, though not easily perceptible, are increasing. The increase has not yet reached catastrophic level. If the imbalance is not corrected, our genetic incapacity in relation to our changing environment will relatively increase. And it is already on increase.

To improve the genetic quality of a population and remove or, at least, minimize the difference between the mean fitness of the population and that of the optimal genotype ("genetic load"), it has been suggested by some biologists like Herman Muller,[32] we should adopt a eugenic policy to allow willing people to use the semen of the persons possessing the desired qualities. Planned reproduction, according to the proponents, stands a better chance to reverse the downward genetic drift by compensating for diseases and debility and also to help the human species progress towards a higher level. Lending partial support to Muller, it has been claimed by Medawar that "a case can be made for saying that a genetical system that attaches great weight to genetic diversity is part of our heritage, and part of the heritage of most other free-living and out-breeding organisms."[33] Strictly on scientific grounds this theory of bio-engineering has been questioned, among others, by Dobzhansky,[34] Dubos,[35] and Mayr.[36] Here is a problem in which again the question of value judgment arises in a very big way. The practice of this sort of bioengineering has undoubtedly yielded very beneficial results in agriculture and animal husbandry. On that analogy, would it be advisable to use it for the improvement of the human race? The question is who is to decide those "desired qualities" necessary for defining "improvement". Certainly, all couples would not volunteer to accept this programme. Even those who would are not likely to take from the future gene banks only the genes of the geniuses. If we are to believe the words of wives, children and parents of many geniuses of the past, are not very encouraging. Certainly we do not like to see our homes and lands crowded with Platos, Buddhas and Confuciuses. Even if, contrary to our expectations, the case turns out to be so in the future, will it be very welcome to the general people of that time? We cannot predict their values and preferences. Assuming we can, should we bind the future generations by our own standards of knowledge and value? It is not merely unwise but also dangerous to allow one to dictate others' values disregarding the latter's culture, time and needs. This shows, once again, the futility of search for values which are not culture-bound.

A somewhat comparable problem we are witnessing in the world just at this moment. For some well known historical and geographical reasons—industrial revolution, imperialism and colonization, uneven distribution of and access to natural resources—the developing countries of Asia are facing an acute population explosion, whereas the developed countries of Europe and North America have no such problem. To control population growth there is an urgent need for industrialization and modernization. The inputs necessary to achieve this twin ends are not readily available in the developing countries. The developed countries, where these are nationally available, cannot actually spare for others. Having met the pressing domestic needs, defense needs, demands for social welfare programmes, etc., the industrialized rich countries are unable to spare even 1 per cent of their GNP as developmental assistance to the poor countries. The rich thinks that the poor is not working hard enough to modernize and industrialize his country, to control the population growth, instead ask for aid and, having received it, does not remain grateful. The poor, on his part, thinks that he is being given much less than what should be, himself getting low prices for his raw and semi-finished materials and, in the absence of bargaining strength, is being obliged to pay high prices for capital goods and technology. In the poor's judgment, to sustain his high wastage-based economy the rich is not discharging his historical responsibility to him. To feed and care the undeserving poor the rich cannot make his own men suffer. The noble idea of evolving a just and new economic order remains on the agenda merely as a talking point of the professionals. And the talk goes on and on. The pessimist forecasts that the conflict between population growth and the economic needs for industrialization cannot be resolved unless the population has reached a catastrophic level.

In this case also we find a conflict of values between the rich countries and the poor ones. Wedded as culturally they are to a set of values associated with industrialization and modernization, the Euro-Americans find it difficult to understand why the Afro-Asians cannot follow their ways—hard work, small family, etc. The latter, rooted primarily in the agricultural mode of production and the related values systems, finds the industrial culture very fast and hot, family life under heavy stress, often breaking up, increased crime and juvenile delinquency, drug addiction, widespread mental illness, alienation and anomie. Admittedly, the picture of the poor countries and the industrialized rich ones I am giving are very general and therefore bound to be incorrect in some respects. But the point remains: before one passes value judgment on other's thought, action and attitude, one is first obliged to *understand* the other against his appropriate cultural context. Otherwise, one's value-prescription for the other is liable to be taken as an unilateral dictation. As we should not bind our descendants by our values through bioengineering or some other ingenious ways, so the economically more evolved should not unilaterally impose his values on the less developed by means of political engineering or some other ingenious ways. Fortunately for us, we, the poor and the rich countries, are contemporaneous and can rationally discuss and perhaps resolve our differences in systems of values and production before they reach the catastrophic level.

REFERENCES

1. Freiedrich Nietzsche, *Beyond Good and Evil,* Vintage Books, Alfred A Knoff, New York, 1966.

2. Immanuel Kant, *Critique of Practical Reason,* Liberal Arts, New York, 1956.
3. Plato, *Republic* (ed.) James Adam, 2nd ed., Cambridge Univ. Press, 1963.
4. G. E. Moore, *Principia Ethica,* Cambridge, Univ. Piers, 1903. See also, W.D. Ross, *Foundations of Ethics,* Clarendon Press, Oxford, 1939.
5. R.B. Perry, *General Theory of Value,* Longmans, Green & Co., London, 1926. —*Realms of Value,* Harvard University Press, Cambridge, Mass., 1954.
6. S. C. Pepper, *The Sources of Value,* University of California Press, Berkeley, 1958.
   —*World Hipotheses,* University of California Press, Berkeley, 1942.
7. Walfgong Kohler, *The Place of Value in a World of Facts,* Liveright Publishing Corpn., New York, 1938.
8. E. C. Tolman, *Purposive Behaviour in Animals and Man,* University of California Press, Berkeley, 1949.
9. Gordon W. Allport, *The Nature of Personality,* Addison-Wesley, Cambridge, Mass, 1950.
10. C. I. Lewis, *All Analysis of Knowledge and Valuation,* Open Court Publishing Co., La Salle, III, 1947.
    —*Values and Imperatives,* Stanford University Press, 1969.
11. Charles Morris, *Paths of Life,* George Braziller, New York, 1956.
    —*Varieties of Human Value,* University of Chicago Press, 1956.
12. See (5) 1954, p. 106.
13. See (6) 1958, p. 2.
14. D.P. Chattopadhyaya, "Bolzano and Frege: a note on ontology" in D.P. Chattopadhyaya and P. K. Sen (eds.), *Logic, Ontology and Action,* Macmillan, New Delhi, 1979.
15. See (11), *Varieties of Human Values,* 1956, p. 119.
16. Rudolf Carnap, *Philosophy and Logical Syntax,* Kegan Paul, Trench, Trubner, London, 1935: A. J. Ayer, *Language, Truth, and Logic,* Oxford University Press, 1936; and Charles L. Stevenson, *Ethics and Language,* Yale University Press, 1944.
17. P.H. Nowell Smith, *Ethics,* Penguin Books, London, 1954; R. M. Hare, *The Language of Morals,* Clarendon Press, Oxford, 1952; Bertrand Russell, *Human Society in Ethics and Politics,* Simon and Schuster, New York, 1955, and Charles L Stevenson, *Facts and Values,* Greenwood Press, Westport, Connecticut, 1963.
18. Julian Huxley, *Knowledge, Morality, and Destiny,* Mentor Book, New York, 1957; *Evolution, the Modern Synthesis* Harper, New York, 1963.
19. J.B.S. Haldane, *The Causes of Evolution.* Harper, London, 1932.
20. Theodosius Dobzhansky, *Evolution, Genetics, and Man,* Wiley, New York, 1955; *Genetics of the Evolutionary Process,* Columbia University Press, New York, 1970.
21. Rene Dubos, *Man Adapting,* Yale University Press, New Haven, 1975.
    —*The Wooing of Earth,* Charles Scibner's Sons. New York, 1980.
22. G. G. Simpson, *Biology and Man,* Harcourt, New York, 1969.
    —*The Meaning of Evolution,* Yale University Press, New Haven, 1967.
23. J. Bronowski, *Ascent of Man,* Little Brown, Boston.
24. See (20) 1970, p. 1.
25. See (22) 1969, pp. 130-48.

26. See (20) 1955, pp. 373-78.
27. C,H. Waddington. *The Ethical Animal,* George Allen and Unwin. London, 1960.
—*Science and Ethics,* George Allen and Unwin, London, 1942.
28. See (18) 1959, pp. 251-83.
29. See (19), 1932, p. 20. See also, his "Man's Evolution: Past and Future" in G. L. Jepser, *Genetics, Paleontology and Evolution,* Princeton University Press, 1948.
30. J. C. Eccles, "Cultural evolution vs. biological evolution," *Zygon,* 8, 1973. See also his, *The Neurophysiological Basis of Mind,* Clarendon Press. Oxford, 1953; and Karl Popper, *Objective Knowledge: An Evolutionary Approach,* Clarendon Press, Oxford, 1972; *The Open University,* Rowman and Littlefield, New Jersey, 1982.
31. See (14), 1979.
32. N.J. Muller, *Out of Night: A Biologist's View of Future,* Vangurd, New York, 1935; *The Uses of the Past,* OUP, New York, 1957; "Genetic progress by voluntarily conducted germinal choice," in G. Wolstenholme, *Man and His Future,* Little Brown, Boston, 1963, pp. 47-62.
33. P. B. Medawar, *The Future of Man,* Basic Books, New York, 1960.
34. Theodosius Dobzhansky, *Mankind Evolving,* Yale University Press. New Havean, 1962.
35. Rene Dubos (21), 1975, pp. 432-40.
36. Ernst Mayr, *Population, Species and Evolution,* Belknap Press of Harvard University Press, Camb., Mass., 1977, pp. 150-61.

BIBLIOGRAPHY

Allen, Garland E., 1975b. *Life Science in the Twentieth Century.* New York: John Wiley & Sons.

Allen, Garland E., *Thomas Hunt Morgan: The Man and His Science.* Princeton: Princeton University Press, 1978.

Bohm, David, *Causality and Chance in Modern Physics,* Routledge & Kegan Paul, London, 1957.

Bohm, David, "Science as Perception –Communication" in Frederick Suppe (ed.), *Structure of Scientific Theories,* University of Illinois Press, 2nd rev. ed., 1977.

Bowler, Peter J., *Evolution, The History of an Idea,* rev. edn., University of California Press, Berkeley.

Broadbeck, May, *Readings in the Philosophy of the Social Sciences,* New York, 1968.

Bronowski, J., *Science and Human Values,* Messner, New York, 1956; *A Sense of the Future,* MIT Press, Camb., Mass., 1977.

Buchdahl, Gerd, *Metaphysics and the Philosophy of Science,* Basil Blackwell, 1969.

Burtt, E.A., *The Metaphysical Foundation of the Modern Science,* Routledge & Kegan Paul, London, 1964. See also E.E.Harris, *The Foundations of Metaphysics in Science,* George Allen & Unwin, London, 1965; and M. Capek, *The Philosophical Impact of Contemporary Physics,* Van Nostrand, Princeton, 1961.

Cassirer, Ernst, *Philosophy of Symbolic Forms,* 3 Vols., New Haven, 1953, 1955, 1957.

Chardin, Pierre Teilhard de, *Man's Place in Nature,* Fontana Books, London, 1971.

Chardin, Pierre Teilhard de, *op. cit.*

Chardin, Pierre Teilhard de, *The Future of Man,* Fontana, London, 1973.

Chardin, Pierre Teilhard de, see (12) above.

Chardin, Pierre Teilhard de, *The Phenomenon of Man,* Fontana Books, London, 1971.

Chattopadhyaya D.P., *Societies and Cultures,* Bharatiya Vidya Bhavan, Bombay, 1973.

——*Interdisciplinary Studies in Science, Technology, Philosophy and Culture,* New Delhi, Project of History of Indian Science, Philosophy and Culture, 1996 and 2004.

——"Experience and Experiment" in *Interdisciplinary Studies in Science, Technology, Philosophy and Culture,* New Delhi, Project of History of Indian Science, Philosophy and Culture, 1996 and 2004.

——*Individuals and Worlds: Essays in Anthropological Rationalism,* New Delhi, Oxford University Press, 1976.

——*Interdisciplinary Studies in Science, Society, Value and Civilizational Dialogue,* Shimla, Indian Institute of Advanced Study, 2004.

——*The Ways of Understanding the Human Past,* New Delhi, Project of History of Indian Science, Philosophy and Culture, 2001 and 2003.

——(ed.) *Philosophical Consciousness and Scientific Knowledge: Conceptual Linkages and Civilizational Background,* Vol. XI, Part 1, New Delhi, Project of History of Indian Science, Philosophy and Culture, 2004.

——(ed.) *Self, Society and Science: Theoretical and Historical Perspectives,* Vol. XI, Part 2, New Delhi, Project of History of Indian Science, Philosophy and Culture, 2004.

——(ed.) *Sri Aurobindo and Karl Marx: Integral Sociology and Dialectical Sociology,* New Delhi, Macmillan and Motilal Banarsidass, 1976 and 1989.

——(ed.) *Societies, Cultures and Ideologies: Analysis and Interpretation,* Mumbai, Bharatiya Vidya Bhavan, 2001.

——and Ravinder Kumar (eds.) *Science, Philosophy and Culture: Multi-disciplinary Explorations,* Part 1, New Delhi, Project of History of Indian Science, Philosophy and Culture, 1999/2005.

——and Ravinder Kumar (eds.) *Science, Philosophy and Culture: Multi-disciplinary Explorations,* Part 2, New Delhi, Project of History of Indian Science, Philosophy and Culture, 1999.

——(ed.) *Science, Philosophy and Culture in Historical Perspective,* New Delhi, Project of History of Indian Science, Philosophy and Culture, 1995 and 2000.

——(ed.) *Some Aspects of India's Philosophical and Scientific Heritage,* New Delhi, Project of History of Indian Science, Philosophy and Culture, 1995/2000.

——*Mathematics, Astronomy and Biology in Indian Tradition: Some Conceptual Preliminaries,* New Delhi, Project of History of Indian Science, Philosophy and Culture, 1995 and 2000.

——(eds.) *Language, Logic and Science in India: Some Conceptual and Hisotircal Perspectives,* New Delhi, Project of History of Indian Science, Philosophy and Culture, 1995 and 2000.

Chomsky, Noam, *Problems of Knowledge and Freedom,* Fontana-Collins, London, 1972.

Collingwood, R. G., *Essay on Metaphysics,* OUP, 1940.

Cuvier, G., 1812. *Recherches sur les assemens fossils des quadrupeds,* etc. 4 vols. Paris: Deterville.

Cuvier, G., 1813, *Essay on the theory of the earth,* Edinburgh.

Darwin, Charles, *The Expression of the Emotions by Man and Animals,* New edn., 1965.

Darwin, Charles, *On the Origin of Species,* (Facsimile of first edition, ed., Ernst Mayr.), Cambridge: Harvard University Press, 1964.

Darwin, Charles, *Natural Selection,* ed., R. C. Stauffer, Cambridge: Cambridge University Press, 1975.

De Vries, H., *Species and Varieties: Their Origin by Mutation,* 2nd ed. Chicago: Open Court Publishing Co., 1906.

De Vries, H., *Variation,* in Seward, 1909.

Dobzhansky, Th., "A Critique of the Species Concept in Biology". *Phil. Sci.* 2:344-355, 1935.

Dobzhansky, Th., *Genetics of the Evolutionary Process.* New York: Columbia University Press, 1970.

Dobzhansky, Th., *Genetics of the Evolutionary Process,* Columbia University Press, New York, 1970.

Dobzhansky, Th., *Mankind Evolving,* Yale University Press, New Haven, 1976.

Dubos, Rene, *Man Adapting,* Yale University Press, New Haven, 1975.

Dubos, Rene, *The Working of Earth,* Charles Scibner's Sons, New York, 1980.

Duhem, Pierre, *Aim and Structure of Physical Theory,* Princeton, 1954.

Dummett, Michael, *Truth and Other Enigmas,* Duckworth, London. 1978.

Eccles, J.C., "Cultural evolution vs. biological evolution," *Zygon,* 8, 1973.

Eccles, J. C., *The Neurophysiological Basis of Mind,* Clarendon

Press, Oxford, 1953; see also, Karl Popper, *Objective Knowledge: An Evolutionary Approach,* Clarendon Press, Oxford, 1972; *The Open University,* Rowman and Littlefield, New Jersey, 1982.

Eccles, J. C., *The Physiology and Nerve Cells,* 1957.

Einstein, Albert, and Leopold Infeld, *The Evolution of Physics,* Cambridge University Press, 1971, pp. 120-21; see also p. 112.

Feigl, Herbert, "The Mental" and the "Physical", in Feigl et al., ed., *Concepts, Theories and the Mind-Body Problem, Minnesota Studies in the Philosophy of Sciences,* Vol.II, Minnesota Press, Minneapolis, 1958.

Feyerabend, P., *Against Method,* New Left Book, London, 1975.

Fisher, R.A. 1930. *The Genetical Theory of Natural Selection.* Oxford: Clarendon Press (Rev. ed. 1958. New York: Dover).

Foerester, Heinz von, ed., *Cybernetics, Circular, Causal, and Feedback Mechanisms in Biological and Social Systems,* 5 vols., New York, 1950-55.

Gamow, George, *The Creation of the Universe,* New York, 1952.

Garland, E. Allen, *Matter, Energy and Life*, 2nd edn., 1970.

Gasparski, W.W. and Klemens Szaniawski, *Praxiology and Decision Theory*, Warszwa, 1975.

Glenn, R. Morrow, "Plato and the rule of law", in Gregory Vlastos (ed.), *Plato II,* 1971.

Goodman, N., *Language of Art,* Harvester Wheatsheaf, Brighton, (1981).

Habermas, Jürgen, *Theory and Practice,* Heinemann, London, 1974, ch.7.

Haldane, J.B.S., "Man's Evolution: Past and Future" in G.L. Jesper, *Genetics, Paleoontology and Evolution,* Princeton University Press, 1948.

Haldane, J.B.S., *The Causes of Evolution,* Harper, London, 1932.

Haldane, J.B.S., *The Marxist Philosophy and the Sciences*, London, 1939.

Haldane, J.B.S., *Everything Has a History,* 1951.

Hampshire, Stuart, *Thought and Action,* Chatto and Windus, London, 1970.

Harre, Rom, (ed.) *Problems of Scientific Revolution,* Macmillan, London, 1975.

Hayek, F. A. von, *The Counter-Revolution of Science,* Free Press, Glencoe, Illinois, 1952.

Held, David, *Introduction to Critical Theory: Horkheimer to Harbermas,* Univ. of California Press, Berkeley, 1980.

Hilgard, E. R., and G. H. Bower (eds.), *Theories of Learning,* 3rd edn.,1966.

Hinde, R.A., *Animal Behaviour: A Synthesis of Ethology and Comparative Psychology,* 1970.

Hole, Frank and Robert F. Heizer, *An Introduction to Prehistoric Archaeology,* Holt, New York, 1965.

Hull, D.L., *Philosophy of Biological Science,* Prentice Hall, Englewood Cliffs, NJ., 1974.

Huxley, Julian, *East and West: Lysenko and World Science,* 1949.

Huxley, Julian, *Essays of a Humanist,* 1954.

Huxley, Julian, *Evolution, the Modern Synthesis,* Harper, New York, 1963.

Huxley, Julian, *Knowledge, Morality and Destiny,* Mentor Book, New York, 1959.

Huxley, Julian, *The Uniqueness of Man,* Chatto & Windus, London, 1941.

Johann Gregor Mendel's [1822-1884] original works, 13 papers, published during 1860s and 1870s had been brought to the notice of professional Botanists by the Dutch Botanist, De Vries, (1848-1935], in the first decade of the 20th century. Thereafter many translations and editions of the book have been brought out. For example, the names and works of W. Bateson, Cambridge, 1902, J. H. Bennett, Edinburgh, 1965, C. Stern and E. R. Sherwood, (1966), London, 1966 may be mentioned in this connection. Secondary literature on the Life and Works of Mendel, among others, due to H. Iltis, F. A. E. Crew, L. C. Dunn and A. H. Sturtevant [*A History of Genetics*), New York, 1965 are very well known.

Kitcher, P., *Vaulting Ambition: Sociobiology and the Quest for Human Knowledge,* MIT Press, Camb., MA., 1985.

Köhler, Walfgong, *The Place of Value in a World of Facts,* Liveright Publishing Corpn., New York, 1938.

Körner, Stephen, *Categorical Frameworks,* Basil Blackwell, Oxford, 1974; Gellner, Ernest, *Cause and Meaning in the Social Sciences,* Routledge & Kegan Paul, London, 1973.

Körner, Stephen, *Experience and Theory,* London, 1966. Division, Government of India, 1970.

Kuhn, Thomas S., *The Structure of Science,* Chicago University Press, Chicago, 1962;

Lamarck, Jean-Baptiste, *Discours d'ouverture* (an VII, an X, an XI, et 1806), ed. A. Giard, *Bull. Sci. de la France et de la Belgique,* Paris, 1907.

Lamarck, Jean-Baptiste, *La biologie: texte inedited*, ed. P. Grasse, *Rev. Scient,* 82-267-276, 1944.

Lamarck, Jean-Baptiste, *Colloque International "Lamarck,"* ed. J. Schiller, Paris: A. Blanchard, 1971.

Lanyon, Wesley N., and W. N. Tavolga (eds.), *Animal Sounds and Communication,* 1960.

Levi-Strauss, Claude, *Savage Mind,* Weidenfeld and Nichlson, London, 1966.

Lewis, C.I. *An Analysis of Knowledge and Valuation,* La Salle, Illinois, 1946.

Lewis, C.I., *Values and Imperatives,* Standard University Press, 1969.

Life Divine, Sri Aurobindo Birth Centenary Library, Volume 19, Pondicherry, 1972.

Lloyd, E.A., *The Structure and Confirmation of Evolutionary Theory,* Princeton University Press, Princeton, NJ., 1994.

Lyell, Charles, 1830-1833, *Principles of Geology, being on Attempt to Explain the Former Changes of the Earth's Surface, by Reference to Causes Now in Operation.* 3 vols. London (Facsimile ed. 1970, Germany: J. Cramer).

Mach, Ernst, *Knowledge and Error,* intr. Erwin N. Hiebert, D. Reidel Publishing Company, Dordrecht, Holland, 1976. See, in particular, ch. XVI, *Presuppositions of Enquiry* and ch. XVII, "Pathways of Enquiry."

Mayr, E., *Animal Species and Evolution,* Cambridge: Harvard University Press, 1963.

Mayr, E., *Systematics and the Origin of Species.* New York, Columbia University Press, 1942.

Mayr, Ernst, *Population, Species and Evolution,* Belknap Press of Harvard University Press, Camb., Mass., 1977.

Medwar, P. B., *The Future of Man,* Basic Books, New York, 1960.

Mendel, J. (Gregor), *Versuche uber Pflanzen-hybridden. Verh, Natur, Vereins, Brunn* 4; 3057, 1866.

Merleau-Ponty, M., *Phenomenology of Perception,* tr., Colin Smith, Routledge & Kegan Paul, London, 1970.

Merleau-Ponty, Maurice, *The Structure of Behavior,* Boston, 1963.

Milne, E. A., *Modern Cosmology and the Christian Idea of God,* Oxford, 1952. See also, Herman Bondi, *Cosmology,* Cambridge University Press, 1960.

Mishra, Umesh, *Conception of Matter,* Allahabad, 1936. Needham tells us that "there are reasons for us to believe that the roots of the theory of *Paramanu* (atoms) go much further back in the history of the Indian thought".

Moore, G. E., *Principia Ethica,* Cambridge, Univ. Press, 1903. See also, Ross, W. D., *Foundations of Ethics*, Clarendon Press, Oxford, 1939.

Morgan, T. H., *Evolution and Adaptation,* New York: Macmillan, 1903.

Morgan, T. H., *A Critique of the Theory of Evolution.* Princeton: Princeton University Press, 1916.

Morgan, T. H., *The Scientific Basis of Evolution,* New York: W.W. Norton, 1932.

Morris, Charles, *Varieties of Human Value,* University of Chicago Press, 1956.

Muller, H. J., *Out of Night: A Biologist's View of Future,* Vangurd, New York, 1935; *The Uses of the Past,* OUP, New York, 1957; "Genetic Progress by voluntarily conducted germinal choice," in G. Wolstenholme, *Man and His Future,* Little Brown, Boston, 1963.

Nozick, R. , *Philosophical Explanations,* Harvard University Press, Camb., MA., 1981.

O'Flaherty, W. D., (ed.), *Karma and Rebirth in Classical Indian Tradition,* University of California Press, Berkeley, CA., 1980.

Passmore, John, *The Perfectibility of Man,* Duckworth, London, 1972; George Woodcock, *Anarchism: A History of Libertarian Ideas and Movements,* Meridian, New York.

Pepper, S. C., *The Sources of Value,* University of California Press, Berkeley, 1958.

Perry, R. B., *General Theory of Value,* Longmans, Green & Co., London, 1926.

Perry, R. B., *General Theory of Value,* Longmans, Green & Co., London, 1954, p. 106.

Perry, R. B., *Realms of Value,* Harvard University Press, Cambridge, Mass., 1954.

Popper, Karl R., *Objective Knowledge: An Evolutionary Approach,* 1972.

Popper, Karl R. *Conjectures, and Refutations: The Growth of Scientific Knowledge,* Routledge & Kegan Paul, London, 1969. See also Charles Singer, *A Short History of Scientific Ideas to 1900,* Oxford. See also, Ernest Hutten, *The Origins of Science,* Allen & Unwin, London, 1962.

Popper, Sir Karl and Sir John Eccles, "Falsifiability and Freedom" in *Reflexive Water: The Basic Concerns of Mankind,* ed. By Fons Elders, Candor Book, 1974.

Prakash, Satya, *Founders of Science in Ancient India,* The Research Institute of Ancient Scientific Studies, New Delhi, 1965.

Prakash, Satya, *Founders of Science in India,* New Delhi, 1965. see also D. M. Bose, *et al.* (eds.), *A Concise History of Science in India,* New Delhi, 1971.

Putnam, Hilary, *Reason, Truth and History,* Cambridge Univ. Press, 1981.

Radnitzky, G. and Anderson, G. (eds.), *Progress and Rationality in Science,* Reidel, Dordrecht, 1978.

Ross, W. D., *The Right and the Good,* Clarendon Press, Oxford, 1930.

Russell, Bertrand, *Human Society in Ethics and Politics,* Simon and Schuster, New York, 1955.

Sarton, George, *A History of Science,* Harvard Univ. Press, 1966.

Sarton, George, *Introduction to the History of Science: From Homer to Omar Kyayyam,* Baltimore, 1927.

Sartre, Jean-Paul, *Critique of Dialectical Reason,* tr., Alan Sheridan Smith, New Left Books, London, 1976. See also, Claude Levi-Strauss, *Structural Anthropology* I, tr. C. Jacobson and B. G. Schoef, M. Layton, Penguin, 1973.

Seal, B.N., *The Positive Sciences of the Ancient Hindus; and Science and Human Progress: Kasami Commemoration Volume* (eds.), D. S. Kothari and D. R. Gadgil, Popular Prakashan, Bombay, 1974.

Sherrington, Charles *Man on His Nature,* 1975.

Simpson, George Gaylord *Biology and Man,* Harcourt, New York, 1969.

Simpson, George Gaylord, *The Major Features of Evolution,* New York: Columbia University Press, 1953.

Simpson, George Gaylord, *Tempo and Mode in Evolution.* New York: Columbia University Press, 1944.

Simpson, George Gaylord, *The Geography of Evolution: Collected Essays.* Philadelphia, Chilton Books, 1965.

Simpson, George Gaylord, *Splendid Isolation.* New Haven: Yale University Press, 1980.

Simpson, George Gaylord, *The Meaning of Evolution,* Yale University, New Haven, 1967.

Skinner, B.F. *Verbal Behavior*, New York, 1957.

Sober, E., *Philosophy of Biology,* West View, Boulder, Co., 1993.

Sri Aurobindo, *The Life Divine,* Birth Centenary Library, Vol. 18.

Stevenson, Charles I., *Facts and Values,* Greenwood Press, Westport, Connecticut, 1963.

Waddington, C. H., *Science and Ethics,* George Allen and Unwin, London, 1942.

Waddington, C. H., *The Ethical Animal,* George Allen and Unwin, London, 1960.

Wallace, Alfred Russell, 1876. *The Geographical Distribution of Animals.* 2 vols. London: Macmillan.

Wallace, Alfred Russell, 1891. *Natural Selection and Tropical Nature,* London: Macmillan.

Wetter, Gustav A, *Dialectical Materialism: A Historical and Systematic Survey of Philosophy in the Soviet Union,* translated by Peter Heath, London, 1958. See also, C. H. Waddington, *Introduction to Modern Genetics,* London, 1950.

Wheeler E. M. Robert, *Archaeology from the Earth,* Clarendon, Oxford, 1954; see also, V. Gordon Childe, *What Happened in History,* Penguin, London, 1960.

Wiggins, D., *Needs, Values, Truth,* Blackwell, Oxford, 1991.

Index